Ethnicity, Integration,
and the Military

Series Introduction

On behalf of the Fellows of the Inter-University Seminar on Armed Forces and Society, I am pleased to welcome *Ethnicity, Integration, and the Military* to our series, IUS Special Editions on Armed Forces and Society. This latest Special Edition contains selections from a number of distinguished scholars of the military and society, including the editors, Henry Dietz, Jerrold Elkin, and Maurice Roumani.

The book contributes to an understanding of three variables that are critical in the political and social life of many Third World nations: ethnicity, integration, and the impact of the military upon both. An innovative analytic schema provides a common framework for the substantive chapters and suggests a means of integrating data from other countries not treated here. In addition, the book contains a wealth of detailed information that will be of great interest to students of the particular nations covered in the volume.

The Inter-University Seminar on Armed Forces and Society, founded in 1960 by Morris Janowitz, is an international and multidisciplinary society of independent scholars interested in the military, military institutions, and their relationships with the broader society. Some of the best scholarly work in these areas is published in the IUS journal, *Armed Forces & Society*. We are a nonpartisan, nonprofit scholarly society and do not accept government subsidies or take positions on political issues as an organization. The activities of the Seminar are underwritten by a core grant from The Ford Foundation and supported by The University of Chicago. We are deeply indebted to these institutions for their continuing support of our activities.

Scholars interested in the work of the Inter-University Seminar, wishing to affiliate as IUS Fellows, or having suggestions for future volumes in this series are invited to contact us at the IUS Secretariat, The University of Chicago, Social Science Building, Box 46, 1129 East 59th Street, Chicago, IL; tel. (312) 702-8694.

John Allen Williams
Loyola University Chicago
IUS Vice Chairman and Executive Director

Ethnicity, Integration, and the Military

EDITED BY

Henry Dietz, Jerrold Elkin, and Maurice Roumani

Routledge
Taylor & Francis Group

NEW YORK AND LONDON

First published 1991 by Westview Press, Inc.

Published 2021 by Routledge
605 Third Avenue, New York, NY 10017
2 Park Square, Milton Park, Abingdon, Oxon OX14 4RN

Routledge is an imprint of the Taylor & Francis Group, an informa business

Library of Congress Cataloging-in-Publication Data
Ethnicity, integration, and the military / edited by Henry Dietz,
  Jerrold Elkin, and Maurice Roumani.
    p. cm.—(IUS special editions on armed forces and society :
no. 3)

    1. Sociology, Military. 2. Ethnicity. 3. Social integration.
I. Dietz, Henry. II. Elkin, Jerrold. III. Roumani, Maurice M.
IV. Series.
U21.5.E84 1991
306.2′7—dc20   91-10233
                                                                   CIP

ISBN 13: 978-0-3670-0392-0 (hbk)
ISBN 13: 978-0-3671-5379-3 (pbk)

Contents

About the Editors and Contributors

Gordon Bennett is an Associate Professor of Government and Asian Studies at the University of Texas in Austin. He specializes in Chinese and Japanese politics and American foreign policy toward East Asia. He is currently conducting research on political legitimation in Asian and comparative perspectives. He has published four books and numerous articles and chapters and has had research grants from NDEA/Fulbright, the Social Science Research Council, and the Ford Foundation.

James Brown is Professor of Political Science at Southern Methodist University. He has written extensively on national security policy and civil-military relations, with materials appearing in *Armed Forces and Society*, the *Journal of Political and Military Sociology*, and *Defense Analysis*. He is associate chairman of the Inter-University Seminar on Armed Forces and Society and has served as Special Assistant to the Deputy Undersecretary of Defense for Planning and Resources at the Department of Defense.

John Sibley Butler is Professor of Sociology at the University of Texas in Austin and holds the Taca Centennial professorship in Liberal Arts. He has written widely in the area of sociology of organizations with special emphasis on race and the military and the sociology of economics. He has a forthcoming book entitled *The Sociology of Economics: Reconstruction of Race Ethnicity and Economics*. At present he is investigating the impact of military service on the attitudes of patriotism and international issues.

Henry Dietz is Associate Professor in the Department of Government at the University of Texas in Austin. His major areas of research involve Latin American politics and comparative urban politics, along with political participation and poverty and politics. He has done extensive field work in Peru among low-income urban groups and has published articles in the *Journal of Political and Military Sociology*, the *American Journal of Political Science*, and *Comparative*

Political Studies as well as chapters in many books. He was a visiting professor at the U.S. Air Force Academy in 1985-86.

Jerrold Elkin holds a JD degree from Columbia University and a PhD from the University of Pennsylvania. He is currently U.S. Assistant Air Attache to India. He formerly served as an analyst with the Defense Intelligence Agency and as an Assistant Professor of Political Science at the U.S. Air Force Academy. His articles have appeared in *Asian Survey, Armed Forces and Society,* and the *Naval War College Review.* He recently was co-author of a chapter on Indian defense policy for *The Defense Policy of Nations: A Comparative Study,* second edition.

Bharat Karnad is the Washington correspondent of the *Hindustan Times.* He completed his MA degree from the University of California at Los Angeles in political science and has written extensively on South Asian security, foreign policy, and defense technology issues.

Maurice Roumani is Senior Lecturer in Politics and the Middle East at Ben-Gurion University of the Negev, where he is also Chairman of the Elyachar Center for the Studies of Sephardi Heritage. He has taught at George Washington University and the University of Maryland and has published widely on immigration to Israel and on Middle Eastern affairs in general. He is the author of *From Immigrant to Citizen: The Contribution of the Army to National Integration in Israel.*

Raju G.C. Thomas is Professor of Political Science at Marquette University and was a Fellow at the Center for International Studies at the Massachusetts Institute of Technology in 1988-1989. He has been a consultant to the Department of Defense and has written a number of books dealing with Indian defense policy, Indian security, and nuclear weapons. His articles have appeared in *World Politics, Orbis, Asian Survey,* and the *Journal of Strategic Studies.*

Claude E. Welch, Jr., is Professor of Political Science at the State University of New York at Buffalo. He has done research on African politics, human rights, and the political role of the armed forces. His books include *No Farewell to Arms?   Military Disengagement from Politics in Africa and*

Latin America, Human Rights and Development in Africa and *Civilian Control of the Military*, along with more than thirty articles and as many contributed chapters. He has served in a variety of administrative posts at SUNY at Buffalo.

Stephen Wright is an Associate Professor of Political Science at Northern Arizona University. He has taught as well at the University of London and the University of Sokoto, Nigeria, and has done extensive research on Nigerian political development and Nigerian foreign policy. He is co-editor of *Africa in World Politics* and author of *Nigeria: The Dilemmas Ahead* and has as well written a variety of articles and chapters dealing with African politics.

1

THE MILITARY AS A VEHICLE FOR SOCIAL INTEGRATION

Henry Dietz, Jerrold Elkin, and Maurice Roumani

The study of the military in multi-ethnic states, especially but by no means exclusively those in the Third World, has become crucial for understanding not only the nature and character of political systems but also ethnic balances and socialization processes. Militaries since World War II and decolonization have come to play a major role in influencing the direction of societal development and in some cases supplanting civilian authorities altogether. Whether the former or the latter, the role of the military in state affairs deserves close scrutiny to answer numerous questions that have been posited as a result of the armies' growing and changing roles and their influence and penetration beyond what were assumed to be their traditional places in the political system.

One of these roles -- the military as a vehicle for social integration -- can best be discussed by formulating three basic questions: How (on a micro level) do young men and women who are conscripted or who enlist in military service become socialized and changed by their experience? Might these individual changes in the aggregate affect the larger civilian society? And (on a macro level) might the civilian and military sectors come to reflect one another over time?

By "integration" we mean here unifying and bringing together what might have been previously separate entities into

1

a whole whose constituent parts have been blended so as to make subsequent separation difficult or impossible (Wriggins, 1966; Ake, 1967). Integration in this sense has for many years been a fundamental component of political and social efforts at nation-building (Emerson, 1960; Bendix, 1964; Connor, 1972; Deutsch and Foltz, 1963; Jacob and Toscano, 1964, 3-11; Geertz, 1963, among many others), and we shall use it here in the same way. It should be noted, however, that integration will be used to examine *intra-group* as well as *societal* cohesion and incorporation.

Some Theoretical Questions

The field of military sociology emerged following World War II, and has produced an immense amount of research (for two bibliographical essays see Lang, 1972; Harries-Jenkins and Moskos, 1971; see also Martin and McCrate, 1984). In a brief review of this literature, Janowitz (1981) identifies three eras of investigation. The first, which he terms "theoretically speculative," emphasized the helpful and/or hindering roles of the military in overall national development and modernization, especially in the newly emerging national-states of Africa, the Middle East, and Asia. The second period Janowitz sees as dominated by empirical investigations, country studies, and institutional case analyses. Finally a third phase synthesizes the two in an interplay between hypotheses and propositions derived from a variety of sources, tested through evidence derived from country studies.

Whatever the periodization, it is not surprising to find that most works do not share a common theoretical framework. In fact, since both Edinger (1963) and Harries-Jenkins and Moskos (1981) lamented the vigorous and consistent use of theoretical frameworks, it is surprising that many still lack any coherent or overt theory at all. Part of the problem for such absence rests with the multitude of subfields and the infinite variety of questions within the general rubric of military-civil relations. Under the general heading of the military's involvement in nation-building, for example, questions abound, and while all in some way are inter-related, theoretical linkages are frequently not spelled out. What contributions and barriers does the military make or pose for political and economic development? Are military governments more or less reforming and efficient than their civilian counterparts? How can a civilian government maintain control over the military? If security is one of the sine qua non of the military, how is security defined, and what can a military do to ensure that a nation is secure?

Such substantive, descriptive, and analytic questions are all of critical interest to scholarly inquiry, but they do not encourage or allow easily generated, common theoretical stances. For the purposes of this volume, however, the editors and the individual contributors assume that any discussion about the role of the military as a vehicle for social integration must take both internal or external characteristics of the military into account as explanatory variables (Van Doorn, 1976). An investigation of the internal characteristics of the military will generally emphasize the organizational aspects of the military, and will concentrate on variables such as professionalism, efficiency, capabilities, structure, and composition of the military itself to explain specific behavior or outcomes. Janowitz (1964) some time ago identified internal characteristics such as organizational format, social recruitment and education, and professional ideology as critical. In contrast, research on external characteristics or linkages will concentrate on the military as part of a larger system by examining the role(s) of the military vis-a-vis civil authority, the overt or covert involvement of the military in politics, and the impact of the military on modernization and development programs. Huntington's (1957) early seminal work cited systemic variables such as the formal position of military institutions in the government and the informal influence of military groups on societal decisions as central for understanding civil-military relations.

The conscious intent of this volume is to combine these two foci. Indeed, given the volume's concentration on integration as its basic theme, any other decision would truncate inquiry, since integration in its complete sense involves change within the institution of the military itself (intra-military integration) as well as change across a society at large, either led, reinforced, or reflected by the actions of the military as an institution (societal integration). A discussion of the source of such change occupies a major place in most of the chapters, since it can come either from outside the military (e.g., from civilian authorities) or from within the military itself. In many of the cases examined here, policies specifically meant to foster or encourage integration within the military have come from external sources such as an executive order or a legislative decree. In others cases, the primary motivation has been intra-military in origin.

The related question of boundaries between civilian and military authority structures is itself a matter of considerable concern in some of the nations studied here, while in others it is not (Huntington, 1957; Harries-Jenkins and Moscos, 1981, 43-76; Albright, 1980). Perhaps a majority of the cases examined involve nations where the distinction between civilian and military authority is quite clearly drawn. But in others, the boundary is either vague or meaningless, especially in nations where the military itself rules and where, as a result, civil-military relations become irrelevant or moot at best. It is in nations such as these (Ethiopia and Nigeria are the clearest instances in this volume) that integration -- a policy that presumably has as one of its goals increased domestic harmony and a lowering of tensions -- may paradoxically provoke the greatest amount of conflict and struggle. Indeed, as a general hypothesis we might offer that conflict engendered by efforts at either intra-military or societal integration will be higher in nations where civilian-military separation is weakest and/or where civilian authority does not have an independent power basis.

How integration is to be obtained and implemented constitutes another source for conflict. Ideally (and given that integration is seen as a desirable goal), integration should be a societally harmonious goal to be advanced through harmonious means. But this utopia can be subverted by any number of means: integration can be interpreted as threatening by a variety of groups in society (including, of course, the military), and its implementation can be undertaken by means ranging from coercion and cooptation to repression or (at an extreme) genocide. Likewise, its implementation can be resisted by an equally wide range of behaviors. For example, civilian authorities can either mandate intra-military or societal integration as a goal, but run up against military opposition to such a goal. Or a civilian regime can mandate integration (either intra-military or societal) through means unacceptable to the military, thereby precipitating a confrontational crisis. Or a military government can decide that force might be the only efficient or satisfactory means for obtaining societal integration, thereby igniting societal conflict. Whatever the specifics, the goal of integration of whatever kind generates at least two essential and analytically distinct (but not separate) sets of questions. First, which groups (the military? ethnically, racially, or culturally identifiable groups?) are the target for integration? Who or what determines that target? Who defines the goals and the means of integration? All such matters emphasize the systemic aspect, but not to the exclusion of intra-military concerns. Secondly, can the military set a goal of integration on its own if it wishes to do so? If so, how does the military act to realize the goal of integration, whether intra-military or societal? Or can it refuse to pursue such a goal if it wishes to? All such matters emphasize the intra-military aspect, but again not exclusively so; these questions obviously raise intra-military concerns as well.

The Systemic Focus

The systemic focus almost inevitably raises the matter of the state, and regardless of whether one takes the side of the neo-Marxists or the non-Marxists (Evans, Rueschemeyer, Skocpol, 1985; Carnoy, 1984), the literature on the state offers some useful leverage for analysis.

Enloe argues that the state is a key element:

ethnicity as a source of individual identity and inter-group conflict (is well researched). But there is a third dimension of ethnicity ...the relationship of ethnic groups and the state ...Ethnic identifications can shape a state, but states more often than is recognized can and do shape ethnic identities and inter-ethnic relations (Enloe, 1980a: 5).

State-ethnic group tensions have been the source of much conflict and tension in many nations for centuries, of course; in general, efforts to increase and to concentrate state power often serve as a spark for ethnic tensions, regardless of whether the state is dominated by civilians or the military (Smith, 1981; Enloe, 1980b:224-230). Furthermore, the state as differential distributor of resources affects intra- and inter-ethnic relations regardless of whether it wishes to or not (Brass 1985:3-10). Horowitz (1985) and others argue that the state can prevent or at least ameliorate ethnically-motivated military intervention through certain policies designed to bring about integration.

Whether the state will be involved in either intra-military or societal integrative policies and conflicts is not thus so much of a question as to how it will become involved. All perspectives on the state (Weberian or Marxist) share a similar concept of the military as a policy instrument of the state. The military is, after all, that arm of the state that represents the state's monopoly over coercion, which is a key element in the definition of the state. And virtually all state theorists agree that the maintenance of order and control are key factors in the formulation and implementation of state policies.

One source of varying goals of integration (e.g., whether integration is designed to maintain the status quo or to reformulate ethnic identity) arises from state concerns about continuing state control and order. Needless to say, the transformation of state goals into military-related policies does not always proceed without incident. The occurrence of a military coup, of course, is the most dramatic and extreme case, since it indicates that the military is no longer an instrument of the state. But disagreements and tensions can emerge in virtually endless variety between the polar opposites of complete acquiescence and adamant refusal. The chapters in this volume provide numerous illustrations of this variety, up to and including coups, and thereby show how what is (or should be) an ultimately pacific or harmonious goal can produce fundamental rupture between two major actors -- the state and the military -- both of whom are critical to the success or failure of national integration.

The Intra-military Focus

If the focus of inquiry concerns the ability of a military to implement policies designed to foster integration (or, more broadly, to carry out state strategies in general) within the institution of the military, then a somewhat different analytic stance is required (see Enloe, 1980b: 219-224). The military as an institution has its own "goal orientation, deliberateness of design, ...boundaries, ...status structure, ...technical system for accomplishing tasks, and ...personnel" (Aldrich and Marsden, 1988: 362), as has any formal organization. But the military is unique in its monopoly over coercive force and its intimate linkages with the state, and therefore any research into intra-military integration must be aware of these linkages, although its emphasis will necessarily be upon goal definition and goal attainment within the institution. Perrow's injunction therefore deserves attention:

> We cannot understand current crises ...without seeing how they were shaped. The present is rooted in the past; no organization (and no person) is free to act as if the situation were de novo and the world a set of discrete opportunities ready to be seized at will. All kinds of structural restraints embedded in the past limit freedom (1986: 158).

Structural restraints include historical and emerging values, existing norms, and the structure of power. The chapters in this volume all discuss the organizational manifestations of these restraints as they influence (and are influenced by) racial and ethnic differences and/or economic class differences.

Any formal organization is subject to a variety of demands from its environment, and is compelled to respond to these demands and to, in one way or another, justify its raison d'etre. One common response to such demands is the constant effort to establish the legitimacy of the institution and of its goals and means to those goals. Environment requires that an organization "establish the legitimacy of its output ...and the legitimacy of its method of operation" (Perrow, 1986), not to mention the legitimacy of its whole existence. The most fundamental output for any military is the rather abstract

notion of national security (Enloe, 1980b, p. 221); legitimizing
that output as well as the methods by which that output is
produced and maintained is an ongoing process that requires
constant attention (Harries-Jenkins and Moskos, 1981: 63-68;
Harries-Jenkins and van Doorn, 1976). Legitimacy, of course,
is a relative term in the sense that what may be legitimate
goals and tactics in one environment or setting would be
illegitimate in another. For example, discriminatory practices
and a non-integrated military might in one setting be viewed as
an appropriate way to establish national security and thereby
legitimacy for the military. In another setting, a vigorous
campaign to integrate the armed forces could be highly
legitimate and legitimizing. And as another example, a military
may enhance its legitimacy through overt attempts to integrate
within itself, thereby setting an example for the rest of the
society in which it exists, or a military may trail behind the
rest of society at large and remain largely non-integrated but
still legitimate. Indeed, in this latter case, a military might
view state-centered policies aimed at national or social
integration as threatening to the military as an institution and
as therefore ipso facto illegitimate.

These theoretical questions offer a general view of some of
the areas of inquiry that are common to all of the cases
studied -- intra-military and systemic concerns, civil-military and
state-military relations, and the impact of such concerns and
relationships on integration and legitimacy. But along with
these theoretical areas are the more substantive questions that
the chapters address, and it is to them that we now turn.

The Military, Ethnicity,
Development, and Integration

Up through the 1960s, the military in the   Third World
was assumed to be an institution of modernization, nation-
building and socialization par excellence (see, for example,
Johnson 1962, 1964; Mullins, 1987: 3-5). A take-over of
civilian power by the military was seen as part and parcel of
"rationalizing" the system and cleansing it from corruption and
anomie. In many ways, the study of the military had given a

certain measure of "idealization" to the role of the military and its achievements.

However, work by many scholars (Huntington, 1968 and Bienin, 1971 are representative) began to question the ability of the military to bring about political and social integration or nation-building. Military governments came under severe criticism when they were discovered to be no different from their civilian counterparts in resisting corruption or in restoring law and order. Such discoveries have been especially notable when the roles of the military in social and ethnic integration are closely examined.

It had also earlier been assumed that the military as a modernizing force would play a significant role in the education and socialization of different ethnic groups by forging integration among them and by creating a homogeneous army (Johnson, 1962). In the cases where universal conscription was implemented, the army was regarded as the "last and best resort" for those who were school dropouts or illiterates. Recruits would be given courses in patriotism and national history. In the process of army service, these recruits would be trained in a particular skill or skills which could be utilized in the civilian sector following discharge. The army would thus contribute directly to the integration of the citizen into the economic, social and political spheres of the state (Janowitz, 1964: 81-82).

This volume demonstrates that such results are not empirically automatic or even likely. In some, the army was unwilling to undertake non-military functions presumed to be the domain of other agencies of the state. Even those armies that undertook to educate, train and socialize their recruits for military service and for civilian life appear to have done so reluctantly and often to have fallen short of the targets they themselves set. Other militaries have acted to control or repress ethnic groups deemed to be important or suspected of disloyalty to the elite or the state on some designed "elite security map" (Enloe, 1980b).

These mixed and ambiguous results raise a number of important questions that are related to ethnic groups and their place in a number of societies. The basic question is whether armies today are capable of contributing to national integration during army service or whether they operate as an arm of the ruling elite with a specific mandate to safeguard internal as well as external state security.

In a major study, Enloe in Ethnic Soldiers (1980b) raised many such questions about the military and its handling of ethnicity and ethnic integration, and it is worthwhile to discuss her substantive findings as well as her analysis. Enloe found ethnicity to possess dynamic qualities that could be used or exploited by ruling elites through their control of (or collaboration with) the military. Such cynicism was especially relevant during colonial rule; in some countries it continued after decolonization. In these cases, the army was found to be perpetuating the interests or the status quo of the particular groups identified with the ruling elite.

In the Third World perhaps more than in Europe, ethnicity became an important and vital component of the forces that shaped the direction and character of society. The military exerted three possible impacts on ethnicity (Enloe, 1980b: 11). First, it could have "no independent effect," and would simply reflect the balance of power among groups in the society. Second, the military could act in a way to dismantle and disarm any ethnic identities so that ethnic divisions would disappear. Third, the military could consciously operate so as to maintain or reinforce ethnic identification.

Elites that govern and control state bureaucracies and agencies are cognizant of the need for a congenial domestic class and ethnic patterns of order for state security. Consequently, "the military may well become an arm of the state elite, charged with erecting and maintaining ...inter-ethnic boundary patterns which make elites feel most secure." (Enloe, 1980b: 14). In such a situation, ethnic groups can be categorized by the military in terms of political reliability and dependability. If, however, each group has a different goal, each may thereby exert an idiosyncratic effect on state security. State elites, therefore, become sensitive to ethnic behavior and are likely to engage in restructuring inter-ethnic relationships to give them a sense of political security. Thus, elites may design an "ethnic security map" that includes geographical dispersion, admission to only certain ranks of state bureaucracies, and assignment to specific ranks within the military.

Enloe (1980a, 1980b) thus confronted several assumptions made in the 1960s when armies were portrayed as prime forces for nation-building. As the collection of the papers in this volume demonstrates, other and perhaps more complex questions now need to be, and can be, asked.

A Framework for Discussion

This book incorporates revised versions of papers written for an Inter-University Seminar on Armed Forces and Society-sponsored panel at the 1986 annual meeting of the American Political Science Association. In addition, several case studies were prepared subsequent to the APSA conference.

The initial provenance of both the panel and book resides in Elkin's research on India. Prior to the Indian army's assault on the Golden Temple in 1984, an act that precipitated a series of mutinies and desertions by Sikh servicemen, Indian defense analysts had repeatedly emphasized the armed forces' signal contribution to national integration. For example, several issues of Sainik Samachar, a semi-official military publication, were devoted to the army's success in generating national, secular perspectives among its personnel. Post-1984 characterizations of the military's integrative role have been more restrained, with attention focused on new structural arrangements such as the Indian Army's establishment of an Institute for National Integration.

The status of minority soldiers within the Indian military initially generated four basic areas for cross-national analysis:

(1) *Reflection of national socio-cultural cleavages in the military.* Are all communal groups represented in the military? Does this representation mirror their percentage of the national population? Does the existence of a variety of communal groups in the military weaken its cohesiveness? If not, why is this the case?

(2) *The military experience as a socializing mechanism.* Is there universal military training? Or is the military a volunteer force, involving a small fraction of the population? Do common military training, corporate life in a highly disciplined

environment, and shared hardships serve to reduce communal identification?

(3) *Purposive social integration efforts within the military.* Are military units fully integrated? Is there trans-community deployment of military personnel? Are promotion and career specialization decisions based on perceived competence rather than communal attachments? Does the military attempt to inculcate national values and perspectives in recruit training and professional military eduction? Are social integration efforts targeted at the families of military personnel? Are retired or separated servicemen used to disseminate national values? If the military has not employed these (or other) coping mechanisms, why is this the case? Does this suggest that the military leadership has failed to internalize the norm of national integration?

(4) *Politico-military implications of social integration efforts.* If the military is in the vanguard of national integration efforts, how has this affected other social institutions? Conversely, will the armed forces likely disintegrate during intense communal conflict as subgroup affiliation takes precedence over military cohesiveness and national unity in the value hierarchy of servicemen? During such conflict, what is the probability of military intervention on behalf of a given subgroup?

The possibility for identifying patterns across disparate geographic regions and levels of development (or for demonstrating that the management of military-ethnic group interaction by central authorities remains an idiosyncratic enterprise) necessitated the inclusion of a number of different cases. But pragmatic considerations (e.g., the impossibility of including all cases of potential interest, difficulties associated with gathering similar data for a large number of cases) forced the editors to reduce those issue areas available for comparative treatment.

In essence, three basic and interrelated variables were selected as fundamental: the military, the state, and a society's ethnic makeup. While each of these has areas of autonomous or (for present purposes) tangential concern, all three overlap one another in ways that can be best illustrated graphically in a Venn diagram (see Figure 1).

Figure 1

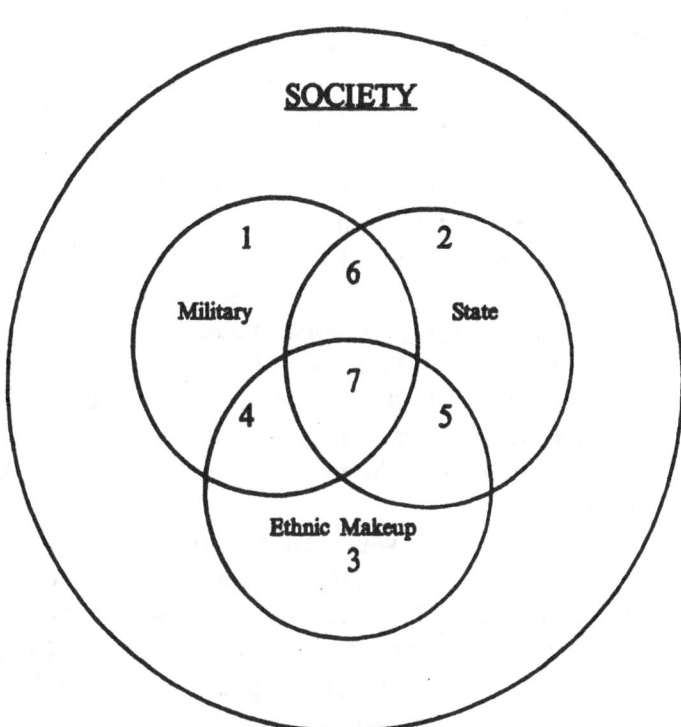

SOCIETY

1

Military

6

2

State

4

7

5

Ethnic Makeup

3

State-Military-Ethnic Interrelationships

<u>Basic Variables</u>

1 - the Military
2 - the State
3 - Ethnic Groups/Characteristics

<u>Interrelationships</u>

4 - Military - Ethnic
5 - State - Ethnic
6 - Military - State

7 - Integration Nexus

As Figure 1 suggests, all three variables immediately and directly impinge on the others. Segment 4, for example, represents state-ethnic relations, and deals with why and how ethnic divisions exist within a society, what the state may or may not do about such divisions (e.g., work to intensify them, ameliorate them, ignore them), and how such divisions may affect state policies. Segment 5 encloses an area that can best be labelled as military-ethnic relations that deal with intra-military efforts to cope with or to resolve ethnic realities of the larger society. These efforts will largely take the form of recruitment, assignment, and promotion procedures as well as conscious efforts by the military to socialize its members vis-a-vis certain state-sanctioned ethnic concerns. Segment 6 thus logically covers military-state relations, including areas such as the role of the military in state security plans (with security understood as internal as well as external), the military as an actor in state ethnic policies, and the degree to which the military and the state do (or do not) converge and resemble or reflect one another.

Segment 7, finally, is what might best be called the integration nexus; it is the coming together of all three major actors simultaneously, and the resulting mix and blend of all three in very considerable part determines the degree to which integration is present or not within a given society and within its military establishment, as well as the degree to which state policies aimed at the integration have been successfully implemented. Integration thus for this volume can be represented by the binary interrelations of all three as well as by the fusion of all three simultaneously. Thus the case studies in the volume address the confluence of these three variables, and we shall at this point offer some brief comments on how the three will be approached and (in highly distilled form) what general conclusions emerge from the several cases.

State-Ethnic Relations

Of all of the areas delineated in Figure 1, state-ethnic relations is undoubtedly the largest and most complex, dealing (at least potentially) as it does with all of the interrelationships between ethnic groups in a society and the state under which they live and are ruled. Yet paradoxically, state-ethnic relations constitute for the purposes of this volume the least directly relevant set of relationships, since they are most usefully conceived of as a contextual variable that provides the setting for the other two (ethnic-military and state-military). Since the focus of this volume concentrates on the military and its role in social integration, most of the chapters provide only a brief overview of state-ethnic relationships in the countries under discussion, and that discussion deals with how the ethnic characteristics of a society (understood throughout as embracing groups differentiated by color, language, and/or religion, thereby covering tribes, races, and nationalities; see Horowitz, 1985: 53) may or may not have an impact on the military as a state-mediated institution. For example, rather than ask simply if ethnic relations in a society are fraught with tension, it is much more to the purpose of the book to ask whether societal ethnic tensions will be reproduced in state-controlled institutions such as the military. Put another way, what can the effect of such an institution be as an intervening variable between ethnicity and social tension (Brass, 1985, Lissak, 1984)? Can social status associated with membership in an ethnic group in the larger society be maintained within such institutions? And to what extent can any institution shaped by uniformed and civilian elites guarantee the maintenance of the existing inter-ethnic distribution of power (Enloe, 1980b: 4-12)? Or can an institution have an independent effect on the notions of ethnic identity in a society?

Virtually all of the authors find that the military has not, in general, played a major or successful role as an integrating device, either within its own ranks or for the larger society. Ethiopia, Nigeria, China, and Israel are instances where this generalization is especially well drawn: Greece, Turkey, and the United States appear to have made somewhat more progress, especially (in the case of the United States) concerning advancement within the military itself.

Yet despite such variations, virtually all of the chapters conclude that militaries tend to reflect the cleavages, stratifications, class relationships, and biases that are present in the larger society in which they exist and from which they draw their primordial influence on all state-run or state-mediated institutions, even if such institutions are determined to act as a change agent within and for the larger society.

Let it be understood, therefore, that ethnicity and its infinitely variable manifestations provides much of the context within which both the state and the military must operate, but that both general and detailed discussions of ethnicity and of state-ethnic relations are beyond the scope of the individual chapters in the volume, and are therefore introduced as they specifically concern state-military and military-ethnic relationships.

Military-Ethnic Relations

In ethnically heterogeneous states, communal affiliation (which can be reinforced by religious and linguistic bonds or geographic concentration) may constitute a prime determinant of military personnel decision. The recruitment, assignment, and promotion of servicemen can be designed to accomplish disparate policy objectives. For example, a civil administration may sustain mono-ethnic control by appointing group members to senior positions in the military hierarchy. Proscribing minority group participation in specified career fields (e.g., intelligence or combat arms) also forms a status confirmation mechanism; such measures may be broadened to include the entire officer corps or, indeed, the military as a whole (Enloe, 1980a: 11-27). Alternatively, ruling ethnic elites may permit outsiders to achieve high military rank, while ensuring that key positions of authority (within and without the defense ministry) remain in the hands of a specific ethnic community. The admission of minority personnel into the military, whether driven by cooptive schemes of manpower requirements, can provide important political socialization opportunities. On the other hand, substantial numbers of armed minority personnel trained in the application of violence ultimately can pose a threat to regime stability. Concern about the reliability of minority troops also may enter the decision calculus of governments employing the military as a vehicle for social integration (for example, in the early years of the U.S. All-

Volunteer Force, a number of observers voiced concern about the overrepresentation of Blacks in the U.S. Army). Nevertheless, nations may still establish objective standards in determining minority access to military positions, recognizing that ties between majority and minority populations will not improve if ethnicity remains a principal criterion for personnel selection and advancement.

Several independent factors may induce minorities to join their state's military forces. Among these are a desire to generate perceptions of trustworthiness to ruling elites, to acquire occupational skills and higher income levels (with a consequent elevation of the group's socio-economic standing), and/or to gain control of the military and, ultimately, all governmental structures. The motivational factor driving a given group may be ascertained (at least theoretically) by identifying its location on the dominant community's threat spectrum. Thus, minorities viewed with profound suspicion by elites may focus their efforts on demonstrating group allegiance to the state.

The case studies in this volume show the continuing relevance of ethnic linkages vis-a-vis the social composition of military forces. Ethnicity remains a matter of salience both for designers of government policy and for minority elements; the latter's sense of separateness may well be heightened by skewed recruitment and career progression patterns favoring the dominant group and its allies.

A number of recurring theses can be discerned in all of the cases. First, they demonstrate that perceptions of communal power and of ethnic group reliability often determine the composition of military forces in ethnically fragmented societies. Second, manpower requirements may exceed in importance the perceived utility of restricting minority access to military employment (i.e., substantial numbers of Blacks were admitted into the US Army during World War I and II to alleviate manpower shortages). Third, military organizations often adopt ethnically differentiated assignment and promotion policies, as illustrated in the Ethiopia, Nigeria, Israel, United States, and China chapters. Fourth, governing elites (in Israel and China, for example) may use the military as an assimilative mechanism, endeavoring to absorb minorities into their own cultural tradition.

These generalizations are consistent with other judgments. A major thesis in Enloe's work, for example, concerns the reflection of inter-ethnic power imbalances in military recruitment, promotion, and assignment politics.

> State elites [employ] ethnic state maps' which trace expectations regarding the political dependability of various ethnic groups. Military policies operationalize state security maps. Official choices in recruiting, promotion, assignments, and field deployments all articulate what state elites are usually restrained from spelling out programmatically (1980b: 15-16).

Such conclusions are not unique to Enloe; Gregory and Ellinwood's (1981) survey of ethnic management policies in South and Southeast Asia and Drysdale's (1982) examination of the Syrian armed forces conclude that military positions of control are given to military officers from ethnic groups deemed safe or secure by state civilian elites.

Governmental willingness to recruit members of politically unreliable communities during high external threat or wartime settings also have been the subject of scholarly inquiry. The need to satisfy manpower requirements led not only to the conscription of blacks and new immigrants in the US, for example, but also the entry of soldiers from African, Asian and Caribbean colonies into the British and French armies. Finally, several authors have investigated how ruling elites employ military service as a socialization vehicle. Moscow's efforts at Russifying minority groups through conscription into its armed forces (dominated by Great Russians) furnishes a useful case in point (Jones, 1981-1982; Jones and Grupp, 1982; Wimbush, 1983).

State-Military Relations

The question to which a military mirrors the society in which it operates is bound to have as many substantive answers as there are cases considered. Most scholars would agree with Guyot (1974), who argues that in order to gain any understanding of how a military related to the larger society of which it is a part, one must first know something about the structures of that society. But while doubtless true, such an admonition does not really constitute or provide guidance.

One more over-arching way to approach the question has been, in recent years, to rephrase it in terms of convergence or divergence, and to ask to what degree civilian and military sectors in a given society are becoming increasingly similar or dissimilar (Lissak, 1984; van Doorn, 1976; Biderman and Sharp, 1968; Moskos, 1971). Lissak's prototypes (1984: 52) are useful here since they allow for institutional linkages between the military and civilian sectors either to be strengthened or weakened as either convergence or divergence takes place.

Since the various cases in this book are disparate and in many ways highly idiosyncratic, they offer some conclusions that may appear to be paradoxical and contradictory. In the first place, where severe ethnic cleavages or discrimination exist in the larger society, they are apt to be mirrored by the military as well and are likely to precipitate difficulties similar to those found in the larger society. This generalization is certainly upheld in the cases of Israel and Nigeria and, perhaps to a lesser extent, Ethiopia and China. The cases of the United States and of Greece and Turkey indicate that societal cleavages can be overcome or at least diminished within a military institution if the larger society is willing to let such cleavages become less important and/or if military professionalization and training have enough strength to overcome such cleavages and/or if the larger society is willing to let professional training take hold. In other words, a military can become an instrument of social integration if, simply put, it wishes to and if society will let it. In those cases where integration has not occurred, neither of these conditions has obtained.

But whether a military mirrors a society or not may not in itself be a reliable indicator of harmony between a military and the society in which it functions (see Decalo, 1976, Chapter 6). Certainly an accurate reflection is not enough by and of itself to prevent military intrusions into the political arena in many countries; after all, only Israel and the United

States have histories of nonintervention, and when intervention has occurred, it has not generally been because of societally-related cleavages. Rather, intervention has had its roots in ideological movements (Ethiopia and Mengistu), foundering civilian governments (Turkey, Greece, China, Nigeria), the military's search for power and/or its greed (Nigeria), or, as is more than likely, some combination of all.

A Methodological Note

The selection of countries and the topics and questions to be confronted in the volume necessitated certain methodological considerations. The first was to identify the parts of the social and political system that were to be compared; the second was to decide upon the number and kinds of countries to be included. The first thus required identification of which political group(s) would comprise the focus of investigation. In this case, the groups under examination included (but were not necessarily confined to) military and civilian elites in several nations. The first also required identifying which political process would be of primary concern. While several were relevant, the political role of the military and the influence of societal/ethnic divisions on civil-military relations are basic to the undertaking. And finally, various social categories had to be identified in order for the comparison to be carried out. For example, having decided that military elites constituted an important political group, what sorts of categories or variables could be selected in order for the comparison to be carried out?

Given the basic question around which the whole volume is wrapped - the role(s) of the military as a vehicle for social integration - the three variables introduced earlier (state, military, and ethnicity) all provide answers to these questions. To discuss recruitment and advancement within the military officer corps, for example, requires not only gathering comparable information on the social backgrounds of this group but also being aware of how the process of recruitment and advancement either parallels or deviates from like processes in the larger society. To describe and evaluate conscious attempts within the military to overcome society's ethnic cleavages and to provide avenues for advancement not found in the larger society requires knowledge of both the society in general and of the military as a group, and of attempts by both to facilitate (or perhaps to inhibit) policies designed to overcome ethnic barriers. Finally, to discuss integration within the military requires distinguishing between integration, (a term that embraces not only thorough-going assimilation but also some rearrangement of status quo peer relationships), and absorption (a term that implies indoctrination or socialization without restructuring existing norms).

The question of country selection is invariably determined by a number of factors, not the least of which is availability of individual scholars who are willing and able to be part of the undertaking. It is seldom indeed that all relevant cases can be included in an edited volume, and neither the individual contributors nor the editors make any claim that all national cases of potential interest have been included in the study. Yet the editors also determined from the start to cast a broad net insofar as the various cases were concerned, thereby consciously adopting the "most different system" approach outlined by Przeworski and Teune (1970).

The advantages of such an approach are considerable (as, it must be noted, the difficulties). Insofar as the former is concerned, comparing nations (or segments thereof) that are in sharp contrast with one another allows for a comparison of "specific situations to exemplify a general reflection on a single process, a single phenomenon" and thereby allows the comparativist "to understand better the essence of the process, to render its definition clearer, to discern its components, and to establish connections between certain contextual variables and its particular manifestation" (Dogan and Pelassy, 1984: 128). A harsh contrast brings into focus those aspects that most prominently characterize the groups under discussion.

Yet countries cannot simply be chosen at random, nor should they be selected simply because they are different. To help guide the selection process, Walton's (1973) idea of a standardized case comparison becomes a useful tool for the collection of original data through systematic and reproducible procedures across cases that are meaningfully comparable

(Walton, 1973: 179-80). Paralleling earlier work by Glaser and Strauss (1967), Walton argues that comparisons carried out through standardized case comparison allows for simultaneous maximization and minimization of both similarities and differences that concern the categories or variables being studied. The ability to hold certain factors constant while manipulating others -- e.g., to focus on specific groups in a variety of nation-states while allowing other contextual factors to vary widely -- may identify communalities between or among the groups in question that could be hidden if only similar systems or contexts were being compared. The bipolar comparisons that could be generated from the cases covered in this volume could, of course, be across similar systems -- for instance, Nigeria with Ethiopia, Greece with Turkey, or the United States with Israel. But some "most different" comparisons are equally intriguing: Nigeria with China, for instance, or Israel with India.

All contributors followed the common framework presented above insofar as it was feasible to do so, given constraints of data and or of national idiosyncracies (i.e., certain factors in one nation that are of vital concern may be less so in another). It is such national variability that makes the "most different" systems comparisons present in the volume especially intriguing. If certain patterns appear across a number of nations, then the volume both through its individual chapters and through its bringing them together will have succeeded in generating and testing descriptive as well as analytic hypotheses.

Notes

1. Other groups, processes, and structures in any society clearly play major roles in the process of societal integration (e.g., overall ethnic heterogeneity, cultural and historical factors, urbanization and industrialization, social class characteristics, to mention only a few), and all of them may directly and indirectly influence the military and its various roles. Identifying and separating these various influences remains an enormously complex and largely undone task.

References

Abrahamsson, Bengt. 1972. Military Professionalization and Political Power. Beverly Hills CA: Sage Publications.

Ake, Claude. 1967. A Theory of Political Integration. Homewood, IL: Dorsey.

Albright, David. 1980. "A Comparative Conceptualization of Civil-Military Relations." World Politics 32, 4 (July), pp. 553-576.

Aldrich, Howard, and Peter Marsden. 1988. "Environments and Organizations," pp. 361-392 in Neil Smelser, ed., Handbook of Sociology. Beverly Hills, CA: Sage.

Baynham, Simon and Richard Snailham. 1983. "Ethiopia." In John Keegan, ed., World Armies. Detroit: Gale, second edition.

Biderman, A.D., and Sharp, L.M. 1968. "The Convergence of Military and Civilian Occupational Structures." American Journal of Sociology 73, 4 (January), pp. 381-399.

Bienen, Henry, ed. 1971. The Military and Modernization. Chicago: Aldine.

Bendix, Reinhard. 1964. Nation-Building and Citizenship. New York: Wiley.

Bradford, Z.B., and F. Brown. 1973. The United States Army in Transition. Beverly Hills, CA: Sage.

Brass, Paul, ed. 1985. Ethnic Groups and the State. Totowa, NJ: Barnes and Noble.

Carnoy, Martin. 1984. The State and Political Theory. Princeton, NJ: Princeton University Press.

Connor, William. 1972. "Nation-Building or Nation-Destroying?" in World Politics 24, 3 (April), pp. 319-355.

Decalo, Samuel. 1976. Coups and Army Rule in Africa. New Haven, CT: Yale University Press.

Deutsch, Karl, and William Foltz, eds. 1963. Nation-Building. New York: Atherton.

Dogan, Henri, and Dominique Pelassy. 1984. How to Compare Nations. New York: Chatham.

Drysdale, Alasdair. 1982. "The Syrian Armed Forces in National Politics: The Role of the Geographic and Ethnic Periphery," pp. 52-76 in Kolkowicz, Roman, and Andrzej Korbonski, eds., Soldiers, Peasants, and Bureaucrats: Civil-Military Relations in Communist and Modernizing Societies. London: George Allen and Unwin.

Edinger, Lewis. 1963. "Military Leaders and Foreign Policy-Making." American Political Science Review 57, 2 (June), pp. 392-405.

Emerson, Rupert. 1960. From Empire to Nation. Boston, MA: Beacon.

Enloe, Cynthia. 1980a. Police, Military, and Ethnicity: Foundations of State Power. New Brunswick, NJ: Transaction Press.

___. 1980b. Ethnic Soldiers. Athens, GA: University of Georgia Press.

___. 1972. Ethnic Conflict and Political Development. Boston, MA: Little, Brown.

Evans, Peter, Dietrich Rueschemeyer, and Theda Skocpol, eds. 1985. Bringing the State Back In. New York: Cambridge University Press.

Geertz, Clifford. 1963. "The Integrative Revolution," pp. 105-157 in Geertz, ed., Old Societies and New States. New York: Free Press.

Glazer, Barney, and Anselm Strauss. 1967. The Discovery of Grounded Theory. Chicago: Aldine.

Gregory, Ann, and DeWitt C. Ellinwood. 1981. "Ethnic Management and Military Recruitment in South and Southeast Asia," pp. 64-119 in Janowitz, Morris, ed., Civil-Military Relations: Regional Perspectives. Beverly Hills, CA: Sage.

Guyot, James F. 1974. "Ethnic Segmentation in Military Organizations: Burma and Malaysia," pp. 27-37 in Kelleher, Catherine M., ed., Political-Military System: Comparative Perspectives. Beverly Hills, CA: Sage.

Harries-Jenkins, Gwyn, and Jacques van Doorn. 1976. The Military and the Problem of Legitimacy. Beverly Hills, CA: Sage Publications for the International Sociological Association.

___. and Charles C. Moskos. 1981. "Armed Forces and Society." Current Sociology 29, 3 (Winter), pp. 1-170.

Herspring, Dale, and Ivan Volgyes. 1977. "The Military as an Agent of Political Socialization in Eastern Europe." Armed Forces and Society 3, 2 (February), pp. 249-270.

Horowitz, Donald. 1985. Ethnic Groups in Conflict. Berkeley, CA: University of California Press.

Huntington, Samuel P. 1957. The Soldier and the State. New York: Vintage Books.

___. 1968. Political Order in Changing Societies. New Haven, CT: Yale University Press.

Jacob, Philip, and James Toscano, eds. 1964. The Integration of Political Communities. Philadelphia, PA: Lippencott.

Janowitz, Morris. 1964. The Military in the Political Development of New Nations. Chicago: University of Chicago Press.

___. 1975. Military Conflict. Beverly Hills, CA: Sage.

Janowitz, Morris. 1981. Civil-Military Relations: Regional Perspectives. Beverly Hills, CA: Sage.

Johnson, John J., ed. 1962. The Role of the Military in Underdeveloped Countries. Princeton, NJ: Princeton University Press.

___. 1964. The Military and Society in Latin America. Stanford, CA: Stanford University Press.

Jones, Ellen. 1981-1982. "Minorities in the Soviet Armed Forces." Comparative Strategy 3, 4, pp. 285-318.

___. and Fred Grupp. 1982. "Political Socialization in the Soviet Military." Armed Forces and Society 8, 3 (Spring), pp. 355-387.

Kourvetaris, George, and Betty Dobratz. 1973. The Social Origins and Political Orientations of Officer Corps in a World Perspective. Denver, CO: University of Denver Social Science Foundation.

Lang, Kurt. 1972. Military Institutions and the Sociology of War. Beverly Hills, CA: Sage.

Levy, Marion. 1966. Modernization and the Structures of Societies. Princeton, NJ: Princeton University Press.

Lissak, Moshe. 1984. "Convergence and Structural Linkages Between Armed Forces and Society," pp. 50-62 in Michel Martin and Ellen McCrate, eds., The Military, Militarism, and the Polity. New York: Free Press.

Martin, Michel, and Ellen McCrate, eds. 1984. The Military, Militarism, and the Polity. New York: Free Press.

Moskos, Charles. 1971. "Armed Forces and American Society: Convergence of Divergence?" in Charles Moskos, ed., The American Enlisted Man: The Rank and File in Today's Military. New York: Russell Sage.

Mullins, A. F. 1987. Born Arming: Development and Military Power in New States. Palo Alto, CA: Stanford University Press.

Perrow, Charles, 1986. Complex Organizations: A Critical Essay. New York: Random House.

Presworski, Adam, and Henry Teune. 1970. The Logic of Comparative Social Inquiry. New York: Wiley.

Smith, Anthony. 1981. The Ethnic Revival. New York: Cambridge University Press.

___. 1982. Theories of Nationalism. New York: Cambridge University Press.

Stockwell, Edward, and Karen Anne Laidlaw. 1981. Third World Development. Chicago: Nelson-Hall.

van Doorn, Jacques. 1976. "The Military and the Crisis of Legitimacy," pp. 17-40 in Gywn Harries-Jenkins and Jacques van Doorn, eds., The Military and the Problem of Legitimacy. Beverly Hills, CA: Sage.

Walton, John. 1973. "Standardized Case Comparison: Observations on Method in Comparative Sociology," pp. 173-188 in Armer, Michael, and Allen Grimshaw, eds., Comparative Social Research: Methodological Problems and Strategies. New York: Wiley.

Wimbush, T. 1983. "The Ethnic Factor in Soviet Armed Forces." Conflict 4, 2-4, pp. 93-179.

Wriggins, Howard. 1966. "National Integration," pp. 181-191 in Myron Weiner, ed., Modernization. New York: Basic Books.

2

THE MILITARY AS A VEHICLE FOR SOCIAL INTEGRATION: THE AFRO-AMERICAN EXPERIENCE AS DATA

John Sibley Butler

The Afro-American military experience is unique in the history of America. This group has participated in all of America's conflicts and wars, making it one of the few ethnic or racial groups that can make this claim about its history. But service to their country has been overshadowed by the realities of slavery and problems of citizenship engendered by the divisive nature of race. In a very real sense, there has been a dialectic of "fear" and "love" guiding the participation of Afro-Americans in the military. Nested within this dialectic, however, is the fact that the military has consistently contributed to the social integration of Afro-Americans into the larger society. The purpose of this chapter is to examine how military institutions have enhanced this integration. We begin by giving a consideration to the theoretical tradition which serves as a guide for our effort. This entails bringing together theoretical ideas of civil-military relations and integrating them with historical data in order to understand the role of the military as a vehicle for social integration.

Theoretical Considerations

The relationship between military institutions and social integration falls under the theoretical tradition of civil-military relations. A major component of this research is concerned with the process by which civilian control of the military is developed and maintained in societies. As history testifies, civilian control of the military has been and remains problematic for numerous societies. It was C.W. Mills who noted that "for the greater part of human history, men have lived under the sword" (Mills, 1956: 171-72) but it is also true that civilians have been able to develop and maintain control of militaries throughout history. The methods by which this is accomplished, and the problems associated with this task, make up the subject matter for students of civil-military relations.

Using historical cases as data, scholars have identified a number of ways by which civilians have been able to maintain a check on military institutions. Constraints built into the constitutions of nation states have been the most frequently used means of civilian control. For example, because of the harsh conditions during the colonial period, the American colonists developed negative attitudes toward a standing military force.   The British forces, housed within cities, treated the colonists as foreigners within their own places of residence. The regular military became a symbol of oppression and external British rule. Thus, the Declaration of Rights in 1774 stated that, "Standing armies are dangerous to Liberty" (Kramer, 1974: 2). When the British were defeated and the constitution constructed, Article II, Section I stated that "the President shall be Commander-in-Chief of the Army and Navy of the United States." Thus, the constitution guarantees a civilian form of government. Other constitutional constraints include legislative investigative powers, budgetary supervision, and investing elected assemblies with the power to declare war and states of emergency (Welch, 1976: 7).

Another method of civil control of the military centers on the importance of ascriptive factors and military service. The classic statement by Mosca that, "the class that bears the lance or holds the musket regularly forces its rule upon the class that handles the spade or pushes the shuttle" captures the importance of these factors (Mosca, 1930: 288). The basic idea is that there should be a convergence between the ascriptive characteristics of top military personnel (e.g. class, ethnicity) and those of the rulers of civilian society. This assures that the class interests of the two spheres are in harmony and decreases the fears of civilian leaders about the possibility of a group overthrowing the government. This relationship observed by Mosca serves as the basis for the recruitment of military organizations in multi-ethnic, segmented societies, where socio-cultural cleavages exist. A consideration of Mosca's observation thus brings us closer to ideas which are germane to the discussion of our major thesis which relates to civil-military relations and the Afro-American experience.

Adekson (1979: 109-125) has developed a framework for examining the maintenance of military organizations in multi-ethnically segmented societies. Based on work in Sub-Saharan Africa, Adekson generates a range of critical questions which emerge when recruitment is considered. Should the military be representative of the larger society? If it is not representative, whom should it represent? What is the effect on the military of the divergences among the people who are citizens of the state? Adekson notes that the answers to such questions depend on the degree to which the political leadership recognizes and accepts the subcultural differences in the society as well as the willingness of the leadership to have the military mirror the society at large. Historical data show that political leaders have utilized three basic kinds of recruitment strategies in segmented societies. They are the ethnic pluralizing approach, the one-ethnic dominant approach and the individual-nationalizing approach (Adekson, 1979).

Under the ethnic-pluralizing approach, groups within the society are the focus for military recruitment. Military organizations recruit from politically relevant groups. Political relevance is measured by a number of variables, including group size. Applying a group calculus, the absolute and relative numbers of members are determined and then recruitment proceeds. Taken to its extreme, the military organizations utilizing this approach constitute an extreme political confederation (Adekson, 1979: 110-11). On the other hand, if a society recruits on the basis of the one-ethnic dominant approach, it starts from the premise that different ethnic interests are not functional for military organizations. The reality of ethnic diversification in the general society is not transferred to the military. Instead, the ethnic group which is dominant is given the "privilege" of serving in the military. This situation is institutionalized and defended mostly on bicultural grounds. The dominant group is depicted as being superior in terms of military skills and political power is also utilized to maintain the position of the dominant group. As Adekson notes, when the dominant military group is not only a numerical minority but also the ruling political elite, the military system may be conceptualized as caste-like in its makeup.

Finally, the individual nationalizing approach accepts the idea that segmented groups within military organizations can be dysfunctional for military organizations. Its major assumption, however, is that individuals who participate do so strictly as individuals and not as representatives of some sub-political or ethnic group. This approach has the effect of minimizing the centrifugal tendencies which multi-ethnic diversity in the military could induce:

> Those who follow this procedure may have another purpose. They may regard a recruitment policy that is nationally and individually based as a means for eradicating in the society as a whole the "evil" which they believe ethnicity represents. In such instances the military becomes a prime instrument for the achievement of civic integration. When national unity is the goal, the size of the military is far less important than is the role it plays in melting and fusing recruits from divergent backgrounds into a single alloy. One's position in the military should reflect and reinforce one's political position. Thus the military participant occupies two overlapping states: that of soldier and of citizen. Not surprisingly, multi-ethnic societies which follow the third approach, particularly if civic integration is their intent, tend to gravitate toward some form of universal and compulsory military service. Obviously, the principle of universal service may be established in nonsegmented societies as well (Adekson, 1979: 11-12).

Thus while the previous approaches solved the problem of civil military relations by exclusion, the individual-nationalizing approach attempts to solve it by inclusion. Put more in the tradition of civil military relations, primordial bonds which are established in the civilian society and which can threaten the stability of the government, are renegotiated or changed so that the individual identifies with the traditions of the nation state as a whole. The Afro-American experience can be grounded in Adekson's work with modifications. Although there were always free blacks, slavery represented a historical reality which greatly influenced the recruitment of blacks to the military. While America, especially after colonial status, tried to stick to a one-ethnic (or race) dominant approach, the realities of manpower shortages forced the recruitment of blacks, both slave and free. Nevertheless, after a conflict was over these individuals were again excluded from the military. In this sense, civil-military relations vis-a-vis Afro-Americans have been influenced by racial factors. It is these factors (racial characteristics) which have dominated the recruitment of the groups to the military. Our historical analysis will show how recruitment policies began with total exclusion and developed into an individual nationalizing approach as discussed by Adekson.

Because Afro-Americans have had such a long and systematic history of military service, exposure to that institution has produced significant benefits for them and the nation. Although the military is an institution which controls the legal means of violence and takes as its major role the protection of nation states, it also serves as an educational institution. One of the by-products of preparing soldiers for the participation in war is the acquisition of skills which can be taken back to civilian life. In a seminal essay, Morris Janowitz examines the importance of this process:

> The armed forces have long been thought of as offering a "second chance" to youngsters from lower-class backgrounds, and even to middle-class youths. A second chance means an opportunity for education and personal development for those who did not have access to appropriate schools and for those who had access but failed. Since its revolutionary origins, the United States has

continuously maintained military and naval forces which have provided these second chances. However, it has only been since the end of World War II that the size of the military establishment has been large enough to supply a significant number of such opportunities (Janowitz, 1971: 167-68).

Janowitz notes that, since 1940, the vast training in engineering, logistics and enterprise within the military produced changes in the occupational and professional structure of the nation. These changes must be viewed in light of the millions of Americans who have served in the military. In a detailed analysis, Janowitz shows the impact of military service on the socioeconomic status of civilians, the consequences of professional training in service for civilians, and the consequences on personal controls and social values which the military exerts. As a composite measurement, when civilians with military service were compared to those without service, their income was much higher. This finding began a tradition of examining the effects of military experience on civilian earnings. Although Janowitz's original model does not give a consideration to race effects, research in later years included this variable. It is this theoretical tradition which will begin to allow us to measure the contribution of the military to the social integration of Afro-Americans in civilian society.

In addition, we must pay attention to the internal dynamics of race relations within the service in order to assess social integration. Charles C. Moskos has noted that what is important is that, when considering minority groups in military organizations, one must build theory which revolves around recruitment, assignment, performance, attitudes towards service life, inter-group relations, and the relationship between armed forces and society (Moskos, 1966: 286). Since Moskos' conceptualization of armed forces and society refers to civil-military relations, his ideas bring together both of our theoretical concerns. He does this by concentrating on both issues of recruitment to the military and how that recruitment affects issues within the civilian sector. The literature on these issues is sufficient to inform our analysis of the contribution of the military to the social integration of Afro-Americans in the larger society.

We begin the analysis by examining issues in civil-military relations vis-a-vis Afro-Americans. Our concern in this section is with questions of Afro-American participation in the military given their dual status as slaves and free people of color. The next section considers the effect of military service on the social integration of Afro-Americans in America.

Race and Civil-Military Relations from the Colonial Period to the Present: A Short Review

The establishment of British colonies in America was intended to provide both markets and raw materials for England. The protection of these interests required military manpower in the form of a militia to defend the colonies against Native Americans and other European powers. Within the colonies, there were both free Afro-Americans in the north and slaves in the south. This situation represented an interesting question of civil-military relations for Euro-Americans during the pre-revolutionary period. During times of conflict, should the colonial militia from each colony depend on Afro-Americans (free or slave) for manpower? In the early days of the colonial experience, Afro-Americans were granted the "privilege" of bearing arms in state militias, even in the south. Consider the following which was passed by the General Assembly of Province of South Carolina in 1703:

XXXIII. Whereas, it is necessary for the safety of this colony in case of actual invasions, to have the assistance of our trusty slaves assist us against our enemies, and it being reasonable that the said slave should be rewarded for the good service they may do us, be it therefore enacted...that if any slave shall, in actual invasion, kill or take one or more of our enemies, and the same shall prove by any white person to be done by him, shall, for his reward, at the charge of the public, have and enjoy his freedom; and the master or owner of such slave shall be paid and satisfied by the public; and if any of said slaves happen to be killed in actual service of province by the enemy, then the master or owner shall be paid and satisfied for him... (Nalty and MacGregor, 1981: 3).

With promises of freedom, Afro-American militia men participated in the defeat of the Yamassee Indians in 1715, but in 1739, a slave revolt occurred, and the colonists adopted a policy of total exclusion for Afro-Americans vis-a-vis military service. The reason for this change in policy was quite logical from the colonists perspective. As noted by a group from the Carolinas, "there must be great caution used, least our slaves when armed might become our masters" (Foner, 1974: 3).

Although exclusion was official policy, Afro-Americans were still armed during times of conflicts with Native Americans. In the French colony of Louisiana, slaves were enlisted to help fight the Chickasaw and Natchez Native Americans. During the four American wars against the French, policies which were designed for black exclusion were overlooked. During these conflicts, Afro-Americans served as soldiers, scouts, wagoners, laborers, and servants (Foner, 1974: 4-5). Thus the reality of manpower shortage, despite the fear of slave revolt, ensured that blacks participated in all conflicts in which the colonist engaged. But the practice of utilizing Afro-Americans during times of crisis and then taking away their arms during times of peace set the stage for race and civil-military relations in years to come. Such a decision was based on the idea in civil-military relations that the group which controls the sword shall have command over those who do not. But in this case, it was also based on the fact that slaves represented property for plantation owners in the south. To send them to war was not a logical economic decision. But overall, within America, there was a fear that arms in the possession of Afro-Americans could be detrimental to the power relations which had developed in the colonies.

It has often been cited that an Afro-American struck the first blow for independence in America. Caught in the cross-fire of the "Boston Massacre", historical documents note the influence of a man named Crispus Attucks:

Monday evening...several soldiers of the 29th regiment were, abusive in the street, with their cutlasses, striking a number of persons. ..A group of citizens, apparently led by a tall, robust man with a dark face, appeared on the scene. There came down a number from Jackson's corner, crying damn, they dare, we are not afraid of them; one of these people, the dark man with a long coward stick, threw himself in, and made a blow at the officer...crying kill the dogs, knock them over. The black man was shot (New York Graphic Society, 1973: 7-8).

Crispus Attucks passed into history, but at the trial of the king's officers that followed, his name was very much alive. John Adams, one of the lawyers for the crown and later the second President of the United States, laid all blame for the revolt on Attucks. He believed that Attucks was one of a motley mob rabble of boys, Negroes, Mullatoes, Irish Teaques and Jack-Tars. Three years later, Adams changed his mind and rallied behind the colonists.

One of the most standard propositions of civil-military relations is that service to country is the highest form of patriotism. But despite the fact that Afro-Americans had served in colonial conflicts before the revolution, when hostilities with England began, they were not welcomed. This shows the powerful element of fear of Afro-Americans bearing arms that had developed during the earlier colonial period. Despite history, and, indeed, the action of Attucks, General George Washington's headquarters issued four orders forbidding Afro-American enlistment in the Continental Army. This action barred the participation of thousands of free Afro-Americans and slaves who were willing to fight on behalf of the colonies.

One of the most interesting historical facts about the relationship between Afro-Americans (slave or free) and the colonial period is that when Washington barred them from service, the British stepped in and tried to recruit them. Early in the conflict, England offered blacks their freedom if they would join the side of The Crown, but the British were unsuccessful because Washington changed his mind due to the realities of manpower. When plans of the British intentions reached him, he issued an order authorizing recruiting officers to accept Afro-Americans who were free. In a letter to the Continental Congress, he wrote in part:

It has been presented to me, that the free Negroes, who have served in the past, are very dissatisfied at being discarded. As it is to be apprehended, that they may seek employment in the Ministerial Army, I have presumed to depart from the resolution respecting them, and have given license for their being enlisted. If this is disapproved by Congress, I will put a stop to it (DoD, 1972: 1).

Because Congress did not disapprove of Washington's actions, over 5,000 Afro-Americans served in the Revolutionary War. As the war progressed, they enlisted freely and many slaves served in the places of their masters (only Georgia and South Carolina refused black enlistment). They were present at the battles of Saratoga, Red Bank, Princeton, Savannah, Monmouth, Bunker Hill, White Plains and Long Island. At Concord and Lexington, Afro-Americans served with distinction and helped propel America toward a successful future. But at the end of the conflict, the conservatism of the constitutional era developed and Afro-Americans were treated without honor. Put differently, the defeat of the British forces that secured Euro-American liberties removed the necessity of allowing Afro-Americans freedom to fight, or their freedom from bondage.

By the time of the Civil War, slavery had become a harsh reality within the American south. Although this war struck at the roots of slave-holder power and although northern free Afro-Americans supported the war, it was not clear whether or not slaves in the south would turn against their masters. The decision not to arm them was naturally grounded in fear:

Fear of armed blacks...caused the Confederacy to ignore a potentially valuable source of manpower.... Although slaves served as military laborers prior to the bombardment of Fort Sumter, the Confederate States rejected any combat role for Negroes, spurning even the services of Louisiana's free black militia, which traced its origins beyond General Jackson's defense of New Orleans in the War of 1812 (Nalty and MacGregor, 1981: 20).

As the war drew to an end, Confederate Major General Cleburne argued that some of the slave population should be set free in return for military service. This argument was based on the belief that the participation of Afro-Americans on the side of the confederacy would bring the support of Britain and France, countries which sympathized with the south but condemned slavery. By the time Confederate leaders endorsed the plan, both slavery and the "southern way of life" were doomed (Nalty and MacGregor, 1981).

The Union Army also spurned Black participation early in the civil war. When northern Afro-Americans responded with enthusiasm to the first call for volunteers, the Secretary of War said bluntly, "This department has no intention to call into service any colored soldiers" (Stillman, 1968: 9). It was not until after the Emancipation Proclamation that blacks were finally allowed to fight the Southerners who had kept them in bondage for so long. Eventually, nearly 180,000 Afro-Americans fought on the union side and were formed into separate army units called "United States Colored Troops." Their units fought in pivotal battles, won fourteen congressional medals of honor, and played major roles in the liberation of Petersburg and Richmond (Hoover, 1968: 270).

After the war was over, the Army reversed its policy regarding Afro-American enlistment, noting that their enlistment was a "peculiarity of the volunteer service." Because of this peculiarity, the War Department noted, black participation had not been authorized for the regular service (Foner, 1974: p. 127). But the Army did create four Afro-American units (there had been 120 during the war); the Ninth and Tenth Cavalry and the Fourteenth and Fifteenth Infantry. These units were assigned to secure the American west, playing major roles in the Indian Wars from 1870 to 1890 (Stillman, 1968: 11).

Although Afro-Americans fought in the Spanish-American War, it was not until World War I that race again became an issue in military manpower. In the civilian sector, there was a continuing decline in race relations. There were sixty-nine lynchings of Afro-Americans in 1915 alone. But when President Wilson announced that he wanted to "Make the world safe for Democracy," more than 200,000 blacks answered the call. At the same time, Frank Park, a U.S. representative from Georgia, introduced a bill which would make it unlawful to appoint blacks to the rank of either noncommissioned or commissioned officers. During the war, the practice of segregated units established systematically in the Civil War continued.

At the end of the war, the Army remained segregated and adopted a policy, the quota system, which kept the number of Afro-Americans proportionate to the total population. By 1940, there were 5,000 men in all-black units, but only five black officers. White officers were assigned to command black troops, a tradition which had developed in the segregated military.

With World War II, America continued its civil-military relations practice of turning to Afro-Americans for manpower. But beginning with this period, the relationship between race and military participation changed significantly. Issues within the civilian sector were tied to the participation of Afro-Americans in the military, and the country was forced to rethink its tradition of the utilization of Afro-American troops.

Disillusionment from the First World War was heavy in the AfroAmerican community. Thus when America turned to this community for manpower, a number of leaders envisioned the experience of the World War I period repeating itself. Military participation and jobs in national defense were tied together, and A. Phillip Randolph called for a march on Washington to end general segregation in American society:

> We call upon you to fight for jobs in National Defense....We call upon you to struggle for the integration of Negroes in the Armed Forces.... This is the hour of crisis. ...To American Negroes, it is the denial of jobs in Governmental defense projects. It is widespread Jim-Crowism in the Armed Forces of the nation (Mier, 1969).

The effectiveness of Randolph's march and protest by other leaders such as Elijah Muhamand and Bayard Rustin would be measured after the war. During the war years, pressures for manpower were tremendous, and the entire country was concerned with defeating Germany. Thus the segregated Army grew as old units were reactivated and new ones were formed to accommodate the more than 900,000 Afro-Americans who served. About three-quarters of these troops were assigned to the quarter-master corps, the engineers, and the transportation corps.

During the period immediately after World War II, the Army began to feel pressure from black and liberal troops to change its segregation policy. In addition to Randolph's earlier actions, the Congress of Racial Equality (CORE) was organized in Chicago, and Gandhi's principles of non-violence found many advocates among Afro-American leaders. It was during this time that the military began the reexamination of the utilization of Afro-American troops. In 1944, Lieutenant General Alvin C. Gillem was charged with establishing a board to study how Afro-Americans could be used more efficiently within the Army's traditional segregated structure. The board interviewed more than 320 white officers, the majority of whom were against integration of the military. A report known as the "Gillem Report" was issued in 1945 and concluded that small Afro-American units together with larger white units were superior to large Afro-American units in all respects (i.e. combat readiness, morale, and moral discipline). The report also suggested that the Army should limit Afro-American recruitment to ten percent of the total Army and continue policies of segregation. Five years later, another Army board headed by Lieutenant General S. J. Chamberlin came to essentially the same conclusion. In 1948, despite the recommendations of these boards, President Truman issued Executive Order 9981 which officially killed the segregation policies of the military. Essentially, this policy noted that there shall be equality of treatment in the Armed Forces without regard to racial characteristics (Davis, 1971: 652).

Truman also established the Fahy Committee which was designed to promote the goal of equal treatment for all military personnel. This committee was instrumental in removing the racial quota system in 1950 and was directing the integration of training camps when the Korean conflict developed (Moskos, 1966: 111).

The need for more combat soldiers was extensive during the Korean conflict. Because of previous protest by Afro-American leaders and the actions of President Truman, Afro-Americans were assigned to integrated combat units for the first time since the revolutionary period. As integration in Korea became standard, it was noted that the fighting abilities of Afro-Americans and Whites were not different (Moskos, 1966: 111).

The aftermath of Korea (sometimes referred to as the Cold

War period) saw a continuing improvement in race relations in the military. Peace-time Army personnel experienced the ongoing implementation of the Army policy of integration. It must be remembered that the word "integration" was truly revolutionary during this period of history. This "revolution" in race relations lead one writer to exclaim that, "The U.S. Army has solved the Negro integration problem still plaguing much of the rest of the nation." (Evans, 1960: 26). Indeed, this period in military race relations has been referred to as the "unbunching of black troops" (Nichols, 1954: 221), a reference to the fact that throughout the country race relations on military bases were better than race relations within the civilian sector.

The Vietnam War saw the highest proportion of Afro-Americans participating than in any other war or conflict in American history. In 1968 alone, 70,000 Afro-American service men were employed, and thousands served during the years that followed. They were over-represented in combat arms and experienced high casualty rates. For example, between 1965 and 1969, they made up 12.6 percent of the soldiers in Vietnam but the total percentage who died hostile deaths was a higher 14.9 percent. Researchers have explained this phenomenon of high death rates by looking at the influence of socio-economic status on military service during this time period. Put differently, this situation occurred not because of race but because of the socio-economic status of Afro-Americans. Regardless of race, lower socio-economic status groups were more likely to serve in Vietnam. Thus, because Afro-Americans are more likely to be of lower socio-economic status than whites, they were more likely to serve and die in that conflict. But one must agree with Lopreato when he notes that, "For my purposes, however, the difference is not of the essence. Whether we victimize one another on the basis of class differences or racial differences, or both, does not gainsay the victimization of the underdog" (Lopreato, 1984: 139).

In terms of civil-military relations, the end of that war saw the development of the All-Volunteer Force, and marked the first time in American history that the military did not try to regulate, after a major conflict, the number of blacks who served.

As the All-Volunteer Force began to blossom, there were few questions raised about the participation of Afro-Americans. But between the years 1973 and 1979, the black enlisted entrants to the Army increased from 19.6% to 36.1%. Although there were no "official" objections from the Department of Defense, two academic scholars raised the question of increasing black participation. A paper entitled "Racial Composition in the All-Volunteer Force: Policy Alternatives" (Janowitz and Moskos, 1974), raised issues of race and the military which had been on the scene for some years. Arguing that the growing percentage of blacks meant that the military would become non-representative of the larger society, they made the following points: (1) that a national fighting force, if it is to enjoy political and democratic legitimacy, must be broadly representative of the population it is established to protect and defend; (2) that units that are over proportionately black will discourage white participation because of fears of "hooliganism" perpetrated by young blacks on whites; and (3) that largely black units, particularly in the combat arms area, will render black troops more highly vulnerable to casualties during times of war. Although there was no official policy developed to stop black enlistment in the All-Volunteer Force, the issues raised by Janowitz and Moskos were debated in the literature (Schexnider and Butler, 1976). The debate served to awaken interest in old issues of civil-military relations and race that had shaped the structure of Afro-American recruitment in past years. Shirley Chisholm, in testimony before the House Armed Services Committee, bridged the gap between the past and the present when she noted:

> All this talk about a volunteer Army being poor and black is not an indication of concern for the black and poor, but rather of the deep fear of the possibility of a Black army. Very few people desire to verbalize the underlying fear and anxiety of a large number of black men trained in the military sense in a nation where racism is rampant. Individuals who are upset over Black power rhetoric really shudder, at the idea of a whole army of black men trained as professional soldiers (Schexnider and Butler, 1976: 425).

Thus after years of loyal participation within the military, beginning with the colonial period, the fear of black participation remained a component part of civil-military relations within America.

The Military as a Vehicle for Social Integration

The Afro-American experience is one of the most interesting studies in human relations in the history of the world. On the one hand, blacks have been utilized as slaves and experienced systematic inequality, but on the other hand, they have shown an unexplained loyalty to their country. Decades before Europeans began migrating to America in large numbers, Afro-Americans had developed patriotism toward a country which had always tried to reject them. Perhaps one can explain this historical anomaly by noting that Afro-Americans have always considered themselves strictly American, with no loyalty to other countries. Unlike their European counterparts, by the late 1800s, they had no sense of the "old country" experience. Historical realities meant that they did not keep a running identity with their relatives from Africa. And, although historians have gone through a great deal of effort to show that elements of African culture influence present Afro-Americans, the fact of the situation is that blacks do not visit their relatives in the old country or even know who their relatives are. With a few exceptions such as Alex Haley, African biological roots have not been traced. Thus, unlike the newcomers from Europe who came to America in waves around the turn of the century and developed dual national identities, Afro-Americans have considered America as their only country.

However, when one examines the development of Afro-American thought, there has always been present the idea that returning to Africa can be an alternative to living in America. It was Martin R. Delaney, a Harvard trained physician and scholar, who visited Liberia in 1859 and originated the phrase "Africa for the Africans" (Cruse, 1967: 5), but Delaney's ideas about returning to Africa were disrupted as he became the first black to be commissioned to field rank in the military by President Lincoln. Put more to the point, blacks have always identified with America as a homeland. It was Frederick Douglass, the great abolitionist and fighter of racism and discrimination who maintained that:

By birth, we are American citizens; by the principles of the Declaration of Independence, we are American citizens; within the meaning of the United States Constitution, we are American citizens; by the facts of history...by the hardships and trials endured, by the courage and fidelity displayed by our ancestors in defending the liberties and in achieving the independence of our land, we are American citizens (Douglass, 1853: 11).

The fact that Douglas (one of the early critics of racism) grounded his argument in service to the country of "our ancestors" as early as 1853 is indicative of the participation of blacks in the military described earlier. Perhaps it is not surprising that the military would become the vehicle for setting the stage for the total transformation of race relations in America.

As noted above, up until the Korean War, America maintained a military which was segregated, a reflection of civilian society. Within military organizations there were limited opportunities for blacks. For example, during the Revolutionary war the typical black soldier was a private who lacked official identity. Often he was carried on the roll as "A Negro man;" or "Negro by name," or "A Negro name not known." Although over 900,000 blacks fought in World War II, approximately three-quarters were assigned to the quarter-master, engineer, and transportation corps. Base facilities were segregated and there was no significant contact between whites and blacks. Total segregation and inequality were not only the rule but the law.

In 1966 Charles C. Moskos, the foremost scholar on race in the military, wrote:

For the man newly entering the armed forces, it is hard to conceive that the military was one of America's most segregated institutions less than two decades ago. For today color barriers at the formal level are absent throughout the military establishment. Equal treatment regardless of race is official policy in such non-duty facilities as swimming pools, chapels, barbershops,   post exchanges, movie theaters and other more formal aspects of the military (Moskos, 1966: 132).

It is also interesting, notes Moskos, that while personnel are often commanded by black superiors, this situation is rarely realized in the civilian sector.

As noted in the previous section, the military was the first institution in America to take significant steps toward equal treatment of personnel, officially ending de jure segregation more than three years in advance of the landmark 1954 school desegregation decision. For millions of Americans this institution served as their first, and perhaps only, truly integrated experience. It has been noted that the post-Korean military, which was preceded by a segregated institution, is an excellent example of the ability of different racial groups to adjust to egalitarian practices, although with some strain (Moskos, 1966). Thus it is an irony that the military, a conservative institution with strong traditions, found itself in the forefront of the experiment of social integration in America.

Despite the movement toward desegregation, research within the "new" military over the last twenty years has documented systematic inequalities. In a major study on promotion rates of enlisted personnel in the Army, it was found that blacks were promoted much more slowly than whites, even after one controlled for education, type of occupation, and test scores (Butler, 1976). Research also found that despite scoring high on tests, blacks were also more likely than whites to be assigned to non-technical occupations (Butler, 1978). Despite these kinds of objective findings, research on black soldiers began to show that they perceived the military institution as being more egalitarian than civilian life (Brink and Harris, 1967). Thus the military as a vehicle for social change, as measured by individual-level perception, was having a much stronger effect within its own confines than within the larger society. Put differently, when the military was compared to the civilian sector vis-a-vis race relations, the former stood in a much more favorable light.

Scholars have devoted a significant amount of time trying to account for the phenomenon of the military being such a rapid vehicle for social change in the country. The literature stresses the unique characteristics of this organization: its separation from the larger society, and its strict hierarchical structure. Although the military is a part of America and its social structure, it has been traditionally a separate entity. Thus the very structure of the military and patterns of human association within it will, to some considerable degree, reflect the social patterns and structure of the civilian society. Nevertheless, the net effect of becoming a part of military organizations is to be separated from one's past life both physically and, to an extent, psychologically. Since the military traditionally was self-contained, all the individuals primary needs were met within it. This separation was consciously promoted by command, especially in the case of new recruits, who are still restricted to the base during basic training. This separation speeds the process of assimilation which results in the recruit's finding the satisfaction of his needs within the military organization.

Once the policy of total racial integration and equality had been established, the discontinuity of military institutions with the civilian sector aided the rapid transformation of race relations. Another factor interacting strongly with the separateness of military society to produce the transformation was the bureaucratic hierarchical power structure of that organization. This structure is highly crystallized. Each office is endowed with specific powers and privileges. Unique to military organizations is the presidential nexus. The president of the country occupies the dual office of chief executive and Commander-in-Chief of the military. Under him the military hierarchy is composed of clearly identifiable grade classes. Rank is indicated by bars and stripes which furnish visible symbols so that all will know how to act toward one another. Because of this highly stratified command structure, once the order for the military to serve as a vehicle for change was given, things move rapidly. Because of the hierarchical structure, which is a pattern of the relationships of command and obedience, decisions regarding race do not have to accommodate the personal desires of military personnel.

Studies have also concentrated on examining the causes of social change within the military. The most celebrated studies concentrate on the relationship between racial contact, which was forced by the structure of that institution, and racial attitudes. Under certain conditions, the more contact individuals have with other racial groups, the less their negative racial attitudes. This is called the "contact" hypothesis. The conditions under which the hypothesis is expected to be true are (a) when an authority positively sanctions interaction, (b) when there are commonly shared goals, (c) when the contact is by equal-status individuals, and (d) when interaction between individuals is cooperative, prolonged, and covers a wide range of activities. The military provides a setting for all of these conditions. Since the publication of The American Soldier, which originally tested the hypothesis (Stouffer et al., 1949), research found that the more racial contact, the less the negative attitudes (Butler and Wilson, 1978).

Thus the unique structure of military institutions is the major explanatory variable in accounting for the military as a vehicle for social change. Indeed, we can say without reservation that it is the only institution where blacks are present, albeit not always at their civilian percentage, in all levels of the rank structure. In corporate or other institutions within the civilian sector, few (and sometimes no) blacks are in higher ranks. Blacks tend to be located in lower management jobs and at the very bottom of civilian occupations. The next question is the effect of military service on the social situation of blacks within the civilian sector.

At a very general level, it is difficult to measure the impact of military service on individuals as they re-enter civilian life. It is quite difficult to directly assess qualities which are developed within the military such as organizational skills, work and communication skills. Scholars have, however, developed a significant research literature which utilizes civilian earnings of veterans, as compared to non-veterans, to access the impact of military service on civilians. Of particular importance is that literature which examines earnings of veterans and non-veterans with controls for race effects.

In this literature, the military is conceptualized as a bridging environment. Bridging environments provide, through work experience, opportunities for movement from a lower to a higher occupation. Browning, Lopreato and Poston (1973) applied this concept to the military and examined the effects of veteran status on civilian income of Afro and Mexican-Americans. As noted in the earlier theoretical section, the military has also been viewed as an educational institution. The Browning, Lopreato and Poston study was able to show that Afro-Americans and Mexican-American veterans received an earning advantage over their non-veteran counterparts. The job training, educational benefits, integration into the living and working arrangements of whites, and the experience of coping with bureaucratic structures of the military were posited as reasons for this advantage. The original findings of Browning and his associates have been repeated in numerous studies which examine the impact of military service on civilian income with controls for race. Like Janowitz's earlier findings on white soldiers, these studies are indicative of the role of the military in the social integration of blacks in America (Lopreato and Poston, 1977; Poston, Segal and Butler, 1984).

Conclusion

The purpose of this chapter has been to examine the impact of military service on the social integration of Afro-Americans. Grounding the analysis in civil-military relations and historical data, we showed that Afro-Americans have always participated, albeit with strain, in the military service. The analysis has shown that military institutions began the integrative process which resulted in the total integration of the Armed forces. This emphasis on the movement toward equality and total participation predated efforts within the civilian sector. An outcome of military participation has been that the earnings of black veterans have been shown to be significantly higher than their non-veteran counterparts. In terms of patterns of recruitment, the American experience has moved from a one ethnic (racial) dominant approach to an individual nationalizing approach. The former excludes minority groups from military participation and the latter stresses individual characteristics of all people in the society, regardless of the groups' membership. In a recent article which appeared in The Atlantic Monthly, Charles C. Moskos (1986) brings together the history of Afro-American "success" within the military officer corp and present day opportunities of soldiers:

> The banquet for black officers of flag rank fairly glittered with stars. Seventy-six black generals and admirals-active, reserve, and retired were being honored at the National Guard Armory in Washington D.C. The date was February 26, 1982. ...The secretary of defense, Casper Weinberger, gave the principal address. ...Some 400,000 blacks serve in an active-duty force of 2.1 million. Most of these men and women serve in the enlisted ranks, many as noncommissioned officers, or NCOs, and an increasing number can be found in the officer corps. Blacks occupy more management positions in the military than they do in business, education, journalism, government, or any other significant sector of American society. The armed services still have race problems, but these are minimal compared with the problems that exist in other institutions, public and private (64).

The phenomenon of Afro-American participation within the military of the United States will continue to be a mixture of success and problems, given the divisive nature of race within Amercian society.

References

Adekson, Bayo. 1979. "Military Organizations in Multi-Ethnically Segmented Societies." Research in Race and Ethnic Relations. Edited by Cora Bagley Marrett and Cheryl Leggon. Greenwich, CT: JAI Press Inc. Volume 1.

Brink, William and Louis Harris. 1967. Black and White: A Study of U.S. Racial Attitudes Today. New York: Simon and Schuster.

Browning, Harley L., Sally Lopreato, and Dudley L. Poston, Jr. 1973. "Income and Veterans Status: Variations Among Mexican Americans, Blacks and Anglos." American Sociological Review 38 (February): 74-85.

Butler, John S. 1976. "Inequqlity in the Military: An Examination of Promotion Time for Black and White Enlisted Men." American Sociological Review 41 (October), pp. 807-818.

Butler, John Sibley and Kenneth L. Wilson. 1978. "The American Soldier Revisited: Race and the Military." Social Science Quarterly 19 (Autumn): 628-638.

Cruse, Harold. 1967. The Crisis of the Negro Intellectual. New York: Random House.

Davis, John P. 1971. The Negro Reference Book. Englewood Cliffs, NJ: Prentice Hall.

Department of Defense, 1972. Report on the Task force on the Administration of Military Justice. Washington, DC: U.S. Government Printing Office.

Douglass, Frederick. 1853. "The Claims of Our Common Cause," in Proceedings of the Colored National Convention (Rochester).

Evans, James C. 1960. "Integration, Differentiation and Refinement." The Negro History Bulletin (April), pp. 151-164.

Foner, Jack D. 1974. Blacks and the Military in American History. New York: Praeger.

Hoover, Dwight W. 1968. Understanding Negro History. Chicago: Quadrangle Books.

Janowitz, Morris. 1971. "Basic Education and Youth Socialization in the Armed Forces," in Roger W. Little, ed., Handbook of Military Institutions. Beverly Hills, CA: Sage Publications.

Janowitz, Morris and Charles C. Moskos, Jr. 1974. "Racial Composition in the All-Volunteer Force: Policy Alternatives." Armed Forces and Society. November, 1974. p. 109-120.

Kraemer, Richard H. 1974. The American Way of War and Peace. New Jersey: General Learning Press.

Lopreato, Joseph. 1984. Human Nature & Biocultural Evolution. Boston, MA: Allen & Unwin.

Lopreato, Sally C. and Dudley L. Poston, Jr. 1977. "Difference in Earnings and Earning Ability Between Black Veterans and Nonveterans in the United States." Social Science Quarterly 57 (March): 750-766.

Mier, August. 1969. Black Protest Through the Twentieth Century. New York: Bobbs-Merrill Company.

Mills, C. Wright. 1956. The Power Elite. London: Oxford University Press.

Mosca, Gaetano. 1930. The Ruling Class. New York: McGraw-Hill.

Moskos, Charles C. 1966. "Racial Integration in the Armed Forces," American Journal of Sociology 72.

___. 1986. "Success Story: Blacks in the Army," The Atlantic Monthly. May, p. 64-72.

Nalty, Bernard and Morris J. MacGregor. 1981. Blacks in the Military: Essential Documents. Wilmington, DE: SR Scholarly Resources, Inc.

New York Graphic Society. 1973, The Black Presence in the Era of the American Revolution. New York: Smithsonian Institute Press.

Nichols, Lee. 1954. Breakthrough on the Color Front. New York: Random House.

Poston, Dudley L., Mady Wechsler Segal, and John Sibley Butler. 1984. "The Influence of Military Service on the Civilian Earning Patterns of Female Veterans: Evidence from the 1980 Census", in Women in the United States Armed Forces. Edited by Nancy Goldman. Chicago: Inter-University Seminar on Armed Forces and Society.

Schexnider, Alvin J. & John Sibley Butler. 1976. "Race and the All Volunteer System," Armed Forces and Society 2: (May): p. 4-21.

Stouffer, Samuel. 1949. The American Soldier. Princeton, NY: Crowell.

Stillman, Richard J. 1968. Integration of the Negro in the U.S. Armed Forces. New York: Frederick Praeger.

Welch, Claude E., Jr. 1976. Civilian Control of the Military. Albany, NY: University of New York Press.

3

THE MILITARY, ETHNICITY, AND INTEGRATION IN ISRAEL REVISITED

Maurice Roumani

The 1960s have seen the emergence of a body of literature on the role of the military in new nations and its effects on nation-building and development. Common among such studies was the belief that these armies in Asia, Africa and Latin America possessed the organizational capability, the technology, and the necessary manpower to achieve modernization and integration. Armies were considered the spearhead of social reforms, and their institution the most suitable framework for promoting equality, acculturation and integration of diverse ethnic groups, often differentiated by race, language, history and religions. The military was generally considered an instrument for forging a more homogeneous society and a stable nation-state.

Empirical data and case studies published in the 1970s and 1980s generally seem to confirm these early assumptions held about the military. However, these studies also raised fundamental questions as to the effectiveness of the role of the military in civilian spheres such as education, occupational skills, and economic mobility (Bienen 1983). These studies also proposed to take a closer look at the role of the military in modernization, integration and ethnic relations (Bienen, 1983; Enloe, 1980a, 1980b).

The Israeli army is a case that repays such examination. For many years, the IDF (Israel Defense Forces) were regarded as a melting pot for exiles. It was an accepted notion that whereas the Israeli society could be accused of some forms of discrimination based on ethnic origin, the IDF was held in high esteem by the nation as an egalitarian institution devoid of any discrimination or favoritism. In effect, the IDF was regarded as an agent of integration where ethnic groups from different social strata and economic classes could meet and spend three years (for men) or two years (for womem) together. General conscription as required by law in Israel guarantees equal rights and responsibilities to all conscripts. Consequently, since its inception, the IDF had been regarded as the only state organization in the country in which ethnic discrimination played no part. The recent drafting by the IDF of the disadvantaged and delinquent youth was considered a further strengthening of its contribution to social change and integration.

The IDF's Role in Immigrant Absorption

Between 1969, when the Camp Marcus study was undertaken, and 1983, when the David Educational Unit was examined, only a handful of articles appeared in journals and newspapers about the IDF's role in immigrant and specifically Oriental absorption. During this period, and particularly after the 1973 Arab-Israeli war, the IDF recognized the significance not only of higher technology in armies but also the importance and implications of deploying masses of soldiers in the field. Consequently, a decision was made to open the ranks and expand recruitment by including populations which previously were deemed unacceptable or inadequate for the purposes of the army. This expansion was not achieved without an intensive debate in the IDF between those who perceived the military as responsible only for the security of the State and those who argued "that it would be a fundamental error not to exploit a national system with unique characteristics which could be used to rehabilitate young people of recruitment age" (Gal 1982: 2-3). The debate according to the IDF presupposed a number of factors:

1) The IDF required combat manpower and hence must endeavor to reach all sectors of the population.

2) The IDF had an interest in solving the social gap and therefore regarded recruitment of disadvantaged youth as a national mission.

3) Since contemporary military systems were highly sophisticated and required skilled personnel, the IDF should offer enrichment courses for those recruits who were dropouts from the educational system.

4) The IDF recognized that the ethno-social problem in Israeli society could undermine the quality of its fighting men and therefore, by raising the standards of its fighting men it would indirectly raise the quality of the civilian population as a whole.

When Rafael Etan was appointed Chief of Staff of the IDF in 1978, he made it his crusade to conscript what became known as the "disadvantaged and weaker population." Their number was estimated to be 3.3 per cent of the annual eligible conscript population (Nativ, 1983). To this end, new education units were established to accomodate these types of new conscripts. The units fell into two categories: one which emphasized education by providing intensive courses in fields which the recruit was found deficient and thus raising his educational level, and the other which emphasized higher motivation and better self-image which was expected to result in the recruit's higher disposition for military service and for the aspiration to reach higher military ranks and occupations. Each unit contained both programs with different emphases.

To validate these assumptions about the IDF's specific role in social change in Israel, two field studies were conducted by the author in 1969 and 1983 respectively. These case studies dealt with Oriental Jews who arrived en masse in Israel from the Middle East and North Africa between 1948 and 1962. The first study focused on a special educational unit called Camp Marcus located near Haifa in the north and dealt with new immigrant recruits, among whom were Holocaust survivors but mostly Oriental Jews, who arrived during those years. The second study examined the David Educational Unit located near Natanya, south of Haifa, and dealt with a different class of population, that of disadvantaged youth, mostly Oriental

Jews who were first or second generation born and educated in the new state of Israel. The purpose of both studies was to determine whether the IDF contributes to the integration of Oriental Jews, small as these groups might be in these specific programs, and if so to what extent, to identify and evaluate the tools used by the IDF in achieving its declared goals, and to examine closely the educational programs and policies of the IDF which were aimed at promoting integration.

Absorption Versus Integration

As the first field research progressed, it became important to clarify whether the IDF was performing an "integrative" or "absorptive" function. The difference between these two terms thus took on a special meaning, since absorption and integration interchangeably (Azarya and Kimmerling, 1980). As we shall see later, the inability to differenciate between these terms had wide implications for the evolution and development of ethnic relations in Israel, at least in its first thirty years of statehood. The question of integration of different ethnic and social groups was raised earlier on by Gaetano Mosca. According to Mosca's theory in The Ruling Class, the polity or the national community is made up of two interrelated components: the social type and the political formula. The social type (the dominant social group or the ruling class) is generally homogenous since it has a common language and common culture, common ideology and geographical origin, as well as an ethnic affinity. The political formula, on the other hand, reflects the cultural, social, economic and political power in the society. Mosca goes further to say that although the dominant social group determines the political formula, the political formula so determined may not be espoused by other social types within the nation (Mosca,1939: 72). When such a situation emerges, polarization among social types becomes inevitable.

Mosca solves this problem by suggesting making the political formula "accessible" to all social types by ensuring that "the complex of belief, moral and philosophical principles that underlies the formula" sink deeply enough into "the consciousness of the more populous and less educated strata of the society" (Mosca, 1939: 107). By minimizing the difference between the "customs, culture and habits of the ruling class and those of the governed class". polarization of elite and mass could be averted. In short, what Mosca advocates is closing the gap through education and socialization, by absorbing the mass into the political formula of the dominant social type.

On closer look, this may prove insufficient, since other social types within the polity (for example, the case of Oriental Jews in Israel in the 1980's) may or may not espouse all aspects of the political formula of the dominant social type. Assuming they do, Mosca warns that once the basic human needs are satisfied, tension is likely to develop between the social types which may prevent their integration because of what has been called "a silken thread of the subtlest fibre - a difference in education, in manners, in social habits, a thread which is not easily broken" (Mosca, 1939: 114).

Clearly, in either case the process of absorption is insufficient for the attainment of integration. The reason is primarily due to the failure on the part of the elite to make the goals of integration meaningful to those beyond the gap in terms of their own beliefs and experiences. In other words, for integration to be successful, it is imperative that the process work in both directions, so that it does not represent only the total and unconditional imposition of the elite's political formula on the mass, with the mass alone being expected to change. As Binder (1964) pointed out, for national integration to be ultimately successful, there must be a mutual and reciprocal "adaptation of beliefs", a predisposition to change in order to achieve "a behavioral and ideological synthesis" of the political formula which will reflect the values and beliefs of both elite and mass" (640). That is to say, the elite must also demonstrate a predisposition to change which may be referred to as receptivity disposition.

The Problem: Aspects of Polarization in Israeli Society

During forty years of statehood, there has emerged in

Israel two schools of thought with regard to national integration. One claims that the ethnic or social gap in Israel is a product of modernization; another claims that the gap is a product of the socialization and absorption processes. Needless to say, the truth lies in the middle of these two schools. The founders of the State of Israel and the pioneers of the twenties for the most part were Ashkenazim, Jews from Eastern Europe and Russia. They were the authors of modern Zionism, the Jewish national movement, which emerged as a response to the Jewish condition in Europe. Although Jewish in its aspiration, the Zionist movement incorporated European culture and ideology, notably Socialism.

Specifically, the first and second Aliyot or waves of immigrants, had a lot in common. They came from a similar Jewish cultural background, spoke for the most part Yiddish, a form of Jewish-Germanic vernacular, later to be supplanted by modern Hebrew, and were imbued with modern nationalism and secularism. They were highly motivated by Zionist ideology which emphasized the redemption of the land by returning to farming and abandoning the trades which Jews traditionally held in the Diaspora. Although the arrival of these pioneers to the country was in waves, they were self-selected immigrants with young and primarily nuclear families. Hence, the new Jewish state which emerged in 1948 was based on these values and ideals, and all other immigrants were expected to adapt to these ideals in order to realize a homogenous society that reflected the Zionist ideals of the ingathering of the exiles.

Soon after the establishment of the state in 1948, Israel was faced with a mass influx of Jewish immigrants from parts of Asia and Africa who became known as Oriental Jews. Between 1948 and 1952, the Jewish population of the new state more than doubled its population, from 716,000 to 1,450,200. (Azarya and Kimmerling, 1980:578). This wave of immigration was different from the previous waves from Europe in many respects. Unlike their Ashkenazi brethren, Oriental Jews were motivated by religious/nationalist fervor to settle in Israel and by the intolerable situation facing them in their host Arab countries after the Jewish State was born. The majority of those who arrived in the country came from traditional Arab and Islamic societies with all that it implied in terms of economic, social and political orientations. Unlike pre-State Oriental immigration, the immigrants of the fifties were composed of extended families, and more often of entire communities which were either mobilized by emissaries from Israel, forced out by conditions in their countries of origin or swept by messianic feelings of redemption to return to the promised land.

Coming from traditional and relatively hostile societies, most of these immigrants arrived with little or no modern education, with traditional skills and trades, and with hardly any financial assets.

The encounter between these two ethnic groups in Israel created what became known as the "clash of cultures." Later on, this clash turned into disparities in the economic, social and political spheres, known as the "social gap" and which posed a threat to the same social fabric which was the product of the egalitarian Zionist ideology. Although by the mid-1970s Ashkenazi Jews became a numerical minority of 45% among the Jewish population in Israel, they emerged as the dominant social group, "western, relatively prosperous, well educated, and substantially in control of the country's institutions," with the "Oriental, relatively poor, unskilled and...underrepresented in the inner power sanctums."

By the late eighties, the number of Oriental Jews increased to consitute between 60 and 65 per cent of the total Jewish population in Israel. The state's policy of absorption did not close the gap between the two ethnic groups despite the socialization and absorption of the new-born first and second generations of Orientals in Israel. On the contrary, these policies appear to have caused social anomie among them. Soon after their arrival in Israel, Oriental Jews realized that their traditional family structure, values and religious practices were not easily transplanted to the new country. Nevertheless, they managed to maintain some semblance of cohesion despite the pressure associated with adapting to the new society and culture. Of course, this had special social costs especially for the young. High rates of illiteracy and school dropouts, truancy, juvenile delinquency and petty crime, became common among certain Oriental groups. Efforts were made by the state to close the gap between the two groups. In some spheres the gap was considered closing; in other aspects it was reported to be widening.

Regardless of the position taken whether the gap was widening or closing, there emerged a consensus among Israeli policy makers that education held the key for upward mobility of the Oriental, thus lessening the threat of "levantinization" of Israel.   For Israelis who harbored such apprehensions, "levantinazion" may well have been equated with "arabization". The dominant Ashkenazi Jews took a paternalistic view of the Oriental immigrants (Smooha, 1983-84), believing that their culture and values had to be subordinated and transformed as rapidly as possible if they were to be integrated into Israeli society and overcome their "backwardness."

It soon, however, became evident that providing equal educational opportunity per se was not sufficient to meet the challenge.   Many Orientals were unable to compete in schools with their Ashkenazi counterparts.   To remedy the situation, the Ministry of Education and Culture instituted a number of programs of preferential treatment for Oriental youth in order to bring their performance up to the Ashkenazi standards. This preferential treatment was not limited to the civilian sector alone.   The IDF was called upon to shoulder the responsibility of integrating new immigrants, particularly those Oriental youth in Israeli society, and to provide them with tools and motivation necessary to overcome backgrounds of deprivation, failure and alienation.

The IDF Programs

About 88 per cent of Israeli male population reaching the age of 18 serve each year in the IDF for a period of three years and about 62 per cent of the female population serve for two years (Nativ, 1983).   Although the number of males and females in the Jewish population is almost identical, exceptions are made for women on the basis of religious or family reasons hence the lower number of female participation.

Considering the draft and the universal conscription to army service in Israel, the IDF almost from its inception was called to introduce compensatory programs for new immigrants, many of whom were Orientals, and later on for marginal youths. For these youth, serving in the IDF was considered to be their "last chance" before they began their adult life and became productive and functional citizens. To this end, the IDF's Center for Advancement of Special Populations (MAKAM) was established in the mid-1970s to absorb and educate these disadvantaged youth. Placement in the program was based primarily on a recruit's KABA score, an anacronym for quality group identification which is derived from pre-induction tests and interviews. It is an aggregate score which consists of four components: intelligence score, level of education, command of Hebrew, and an index of motivation to serve (Gal, 1986: 79). This last index is determined by looking at such factors as criminal records, juvenile delinquency, psychological problems, parents' level of education and overcrowded living conditions. The range of KABA scores is 41-56 distributed along a standard bell curve with 49 representing a median score. The IDF classifies recruits according to the following criteria: KABA 41-46 (low quality); 47-50 (medium quality); and 51-56 (high quality).

Those recruits falling in the KABA 41-42 range (known as RAM 4-B) have been provided with compensatory education since 1974. Currently there are somewhat fewer than 200 recruits per year falling in this category. Three educational facilities exist to meet the needs of these and other low quality disadvantaged recruits: Havat Hashomer, the David Educational Unit, and the Marcus School.

The Havat Hashomer program is for those with the lowest KABA scores and those with KABA scores of 43-46 who have social, medical or psychological problems. The recruits report to Havat Hashomer immediately after induction for a 12-18 week intensive and personalized program. The David Educational Unit is for recruits in the 43-46 category without social or medical problems, but with serious educational deficiencies. Their course work primarily addresses these deficiencies. Soldiers at both schools must pass a Hebrew proficiency examination before continuing on to their assigned units. Those who fail are sent to Haifa to the Marcus

School for twelve more weeks of Hebrew and other basic
educational courses according to need.

The Camp Marcus School:
Programs and Evaluation

The Marcus School has undergone several stages since it
opened its doors to the influx of Holocaust survivors and
many of the new immigrants from the Middle East and North
Africa.   Until 1956, it was a training ground for elementary
Hebrew.   By 1962, it provided different levels of Hebrew
language training. After 1962, these programs of Hebrew were
transferred to Command Schools, and Camp Marcus became
the last resort for obtaining an elementary school certificate.

During these years, the Camp Marcus School, in addition
to having enrolled those soldiers who failed to demonstrate
proficiency in Hebrew at the end of their graduation from
Havat Hashomer or David Unit Schooling, also accepted
soldier   volunteers with a KABA score of 46.   This latter
group sought to improve its knowledge of Hebrew or to
improve in other areas of primary or secondary education.
Many of these volunteers (almost half) decided to extend their
military service and continue at Marcus where they could enroll
in a college preparatory program.   Female soldiers might also
attend Marcus on a volunteer basis during their last four
months of service. It should be noted that women conscripts
must have a minimum KABA of 46 to be admitted; thus they
were not found in other schools designed for disadvantaged
youth.   Finally, new immigrants (those who have been in Israel
for the one year required to be eligible for military service and
were non-Hebrew speakers) were given a three month course at
Marcus prior to basic training.   While the course emphasized
Hebrew instruction, it also featured the history of Israel and
Zionism combined with trips throughout the country which
served to familiarize the new recruit with the geography and
history of Israel.

In the educational and literacy programs, the IDF
ensured that the conscript's level of Hebrew was adequate for
the purpose of the IDF's activities.   Those who had not

completed elementary education were required to enroll in an intensive course prior to their discharge in order to complete their elemenatry school requirements (Roumani, 1985).

In vocational training, those Orientals who attended these schools were given skills and knowledge which were minimally required to serve the IDF's needs. For the most part this meant filling the lowest occupational rungs of the IDF.

Other programs included the strengthening of the new recruit's identification with the State and the Land of Israel. In this regard the IDF succeeded in inculcating in those Orientals a sense of belonging to the Land by transforming their religious nationalism into effective patriotism. However, the IDF's success in creating in the Orientals a sense of solidarity with the State and the Land was counterbalanced by its failure to create a parallel solidarity between them and the Ashkenazim which would endure beyond military service. It was found that after military service, Orientals and Ashkenazim continued to share their identification with the Land and with the symbols of the State, and their preparedness to defend it at all costs. However, this solidarity was not extended among ethnic communities, and therefore integration was vertical rather than horizontal.

With regard to other subjects, content analysis of courses in history and Hebrew language administered in the IDF between 1948 and the early seventies showed lack of sensitivity and information about the Orientals and their culture and instead portrayed a social situation of polarization between the two ethnic communities by emphasizing the "backwardness" of the East and the "enlightenment" of the West. This picture was a reflection of the civilian society into the army where the negative self-image of the Orientals was perpetuated. Whenever the IDF's textbooks described a cultural or social pattern, it invariably reflected those of the predominant social type. For instance, when a family situation was described, it was not the Oriental type of family structure but only that of an Ashkenazi "egalitarian" and nuclear family which was used as an example of an ideal and "correct" family relationship. When a model Sabbath was described, it was not the traditional day devoted to prayer, religious observance and family gathering which was familiar to the Oriental, but was solely depicted as a secular day of leisure which was commonly observed by many Ashkenazim (only 30 per cent of Jews in Israel identify themselves as religious) and which was held as an ideal to be emulated. Moreover, the prayer book which the IDF provided for those who declared to be religious follows the Ashkenazi version of liturgical traditions, though the IDF prided itself for years for having composed a new prayer book common to all ethnic groups.

The study of the Camp Marcus School found that the IDF programs, in effect, re-established, strengthened and perpetuated the Oriental/Ashkenazi dichotomy which existed in the civilian society, and contributed only in a very small measure to the social and economic mobility of those Orientals who attended these programs. It also confirmed that the IDF's programs were absorptive and not integrative as was the case in the civilian population.

The David Educational Unit

In 1983, of the 5200 recruits annually who met the criteria for course participation, 4200 actually reached in 1983 the IDF educational units. The David Educational unit is an army installation named after David Ben-Gurion, the founder and the first Prime Minister of the new State. The David unit holds four sessions a year for 200 to 230 participants at one time. According to the IDF the criteria for admittance includes low educational level of parents and recruit, crowded housing, low level of literacy, inability to abstract, and general cognitive problems.

The IDF also assumes that low motivation for military service results "from numerous experiences of failures, lack of appreciation of theoretical and abstract concepts, alienation from the school curriculum and distrust in any authority figure." Therefore, the main objectives of the curriculum was designed to increase motivation and self-confidence in the recruits, improve their self-image and fill the gaps in education.

After their basic training in the regular army units, these special conscripts who met the KABA requirement score of 43-46 and the "Personality Profile" which determines the level of their motivation are sent to the David unit for a period of seven to eight weeks.

The staff at the unit consists of company CO's and deputy CO's, a staff of instructors of female soldiers, teaching coordinators and project commanding officer. The curriculum varied from one course to another and included activities such as study-hours, trips, and social and cultural events.

The David Educational Unit: Program Evaluation

The IDF claimed that the primary objective of the course was to enhance motivation for positive service in its ranks and only secondarily to fill the gap in education.

At the beginning of the course, the IDF representative informed the participants that the course was aimed at completing their education. This aroused in the participants false anticipations as to the contents of the curriculum. They were convinced that the intention was primarily to fill the gaps in their basic education.

At the end of the course, they were asked to assess to what degree the knowledge acquired in the course would contribute to their future. 55% claimed that the course had made no contribution at all, 15% said it had helped them in many areas; 30% said that in general it had been helpful but expressed several reservations, such as: a) the content had been mainly military training rather than vocational skills, and thus was considered by them as unsuitable for civilian life; b) minimal emphasis was placed on subjects of basic education, particularly English and arithmetic which were not studied at all even though the soldiers considered them to be of vital importance to themselves; c) other subjects were taught at a low level and most of the material was already familiar to them; d) study periods were not continuous but were interrupted often to include disciplinary roll calls which affected concentration on studies.

The framework of the course was found to be asymmetrical. The IDF aspired at "improving the reciprocal relations between the soldier and the society" by social and vocational mobility. In actual fact, the course emphasized enhancing motivation for service in combat units and guaranteeing continued and uninterrupted military service. A survey has shown that soldiers tended to plan their economic future on the basis of skills they acquired in the course of their military service. It was their expectation that the three years spent in the IDF would assist them in acquiring these skills. But the placement of these soldiers in administrative and service units was not furthering the proclaimed objectives of the IDF, i.e. the advancement of a "weaker population," since it had been perpetuating a situation whereby this population continued to serve in low status occupations.

History and Civics

One of the IDF instruction sheets put out by the Commanding Officer of the Unit described the aim of the course as providing good citizenship training and identification with the State. In evaluating the content of the course the author had the following three objectives in mind:

1) To examine the way in which the course treated subjects such as history and Oriental Jewish heritage which were considered one of the tools for promoting integration among the various ethnic groups. It was expected that these subjects would correct the ethnic stereotypes which emerged in Israel and had become part of Israeli culture since the early fifties. The way to enhance the soldiers' self-image was to ensure a positive approach to their heritage. The attitude of the State to the subject had been paternalistic from the beginning and ignored much of the contents of the Oriental Jewish heritage. Consequently, it perceived Oriental culture as separate from and inferior to that of "Israeli culture" as formed by the state elite. This picture has been gradually changing since 1976 when by act of Parliament, a Center for the Integration of Oriental Jewish Heritage was established in the Ministry of Education and Culture to promote and support research and study of Oriental Jewish Heritage in all Israeli universities and other institutions.

2) To examine how the course dealt with Israel's ethnic

and social problem and how it acted to enhance national identification and integration in Israeli society.

3) To analyze the pedagogic methods of teaching.

The picture which emerged from the study showed that the curriculum for the study of various Jewish communities accentuated their cultural difference by emphasizing the progress made by Ashkenazi Jewry and the backwardness of Oriental communities. The material collected and presented by the teachers neglected to emphasize the common elements which Jewish communities shared; the selection of the material was superficial and analysis of the historical processes were lacking. Furthermore, it was stereotypical and dated. The following is an example of a passage of a portrait of Moroccan Jews:

> They brought with them a lifestyle and outlook which were incomprehensible to many Israelis. Their educational standard was low and they suffered from poverty and unemployment. Most of the immigrants had a temperament which could cause clashes with the neighbors from different cultural backgrounds. The Moroccan Jews were not always welcomed in Israel... Attempts to assimilate and attain equality stems from a sense of inferiority and shame. Parents lack education and often behave in a primitive fashion. Education ceases at an early age because of economic reasons and because of their outlook on life.

In contrast to this portrait of Moroccan Jews, a much more favorable picture was painted of Russian Jewry who in Czarist times allegedly "exerted a decisive influence on all Jewish Diasporas and had a rich cultural and spiritual life." Recent immigrants who came from Russia, it was noted, were concentrated "in liberal and academic professions (and) have been absorbed into senior positions in Israel."

According to the staff in charge of curriculum and instruction, the conscripts of this course suffered from a low self-image as a result of severe stereotyping, and although the aims of the course were to improve their self-image, the course had the opposite effect.

Treatment of the Social Gap

Since 95 per cent of the participants in this course were of Oriental origin, it was important to examine the implications of the ethnic problems in the course. The assumption was that suitable discussion of the ethnic problem in Israel could reduce the sense of alienation, enhance involvement in problems of vital concern to the country, and contribute to enhancing motivation for social commitment in general and military service in particular.

At the end of the course, soldiers were asked about inter-ethnic relations and their answers point at considerable alienation and self-hatred. The following is a sample of their answers:

> The Ashkenazim came to Israel first and that is why I am discriminated against. Everything is given to them first.

> The Ashkenazi hate Moroccans. I have never met Ashkenazim. They say that Ashkenazim have better brains than Moroccans, but I don't think so.

> We changed our surname to an Ashkenazi name. It was Kamal before and now its...and the fact that I look Ashkenazi helps me.

> Being a cook is enough for me. I don't know any Ashkenazi. I only go around with Moroccans. I don't mix with Ashkenazi. They live by themselves somewhere. I have got no Ashkenazi friends. Only Moroccans.

> I want to get on in life but you know the Moroccans, they haven't got the head for it. The Ashkenazi do everthing to get high positions. We don't because we haven't got the head for it. We don't want to make the effort.

These stereotypical views expressed by soldiers were not discussed or examined during the course. Teachers of the course believed that such a discussion was out of place in the IDF and it might lead to "an explosion." But when teachers were asked similar questions, it was obvious that they too were entrapped by stereotypical concepts common to Israeli society. Some examples may illustrate the point:

> We have to fight against the stereotypes even if there is some truth in them. They are not capable of doing anything. Some of them are real criminals. Most of them cannot read or write.

> Their parents are always telling them how well off they were in Morocco. And they came to Israel with nothing. Their parents simply never worked in the occupations we need here in Israel and they had to move up by their own efforts.

> It is the mentality of their parents. They are uneducated and can scarcely read and write. They want their children to help them. The father beat the mother and the children see it all. The house is neglected, old.

> I have a different mentality. Where I come from everyone wants to go into the army. My mother spreads cloth on the table, cleans up. There are sheets on beds. My parents are well mannered. I have values of mutual aid and they are different from their values.

> The gap starts with the parents. They have a different scale of values, different manners and mentality. Our society is more progressive. It's a question of Western culture against Eastern culture. Our culture is more progressive, more rational, achievement oriented and progressive. Their culture is at a more backward stage. I look at them from above. My culture has passed theirs.

These statements are self explanatory. They showed an attitude of paternalism, superiority and (to a certain extent) contempt. It is obvious that teachers who averaged 19-20 years of age were inexperienced, ill-trained and definitely innocent. Although they showed dedication, motivation and hard work, most of them have come from prosperous middle-class neighborhoods and had no prior contacts with this type of population.

Two main conclusions were drawn from the research findings at the David Educational Unit:

1) The IDF's objective of recruiting young people from

special disadvantaged and weak populations in order to improve their adaptation to military service in general and to volunteer service in combat units in particular, met with only partial success.

2)   At the same time, the expectations of the recruits that they would acquire occupational skills or learn arithmetic and English, were not fulfilled.

Thus the absence of reciprocity between the IDF and the majority of this population remained unchanged and reflected the picture prevalent in the civilian society.

The IDF and Ethnicity

The military in Israel has been considered a "melting pot" where no ethnic distinction is made and each conscript has the opportunity more than anywhere else in the society to prove himself and his capability. It is an organization where each individual is assured equality of rights and obligations. In fact, the three years of general conscription, similar to the crucial years a youngster in the USA spends in college, guarantees a world of common values and common experiences for all ethnic groups. It is an intensive experience which draws all groups together. In that case, it is assumed that problems of ethnicity do not arise in the IDF. Thus, the IDF is believed to contribute to the lessening of ethnic rifts in the society.

The findings of the two case studies, however limited these may be, pose the critical question of why the IDF has not been more successful in the social and economic mobilization of ethnic groups and their integration into the society. Specifically, were the IDF programs aimed at effectively dealing with the socio-economic problems of ethnicity, or was the IDF merely a representative, or another arm of the State's elite? Why did not the programs of the IDF contribute to the abatement of ethnicity in the society, now that two generations of Israeli born Ashkenazi and Orientals have completed their military duty? To answer these questions, we found Enloe's writings (1980a, 1980b) to be instructive and of relevance to our studies.

In her writings, Enloe sees ethnicity as a dynamic force

which is "open to changing collective definitions and fluctuating emotional intensities." (1980a: 2). According to this theory, the state elite in any plural society has a clear idea "of which ethnic groups are most reliable" and thus invariably "the ethnic group with which the members of the State controlling elite themsleves identify." (Enloe, 1980a: 15). Thus in military planning, both the state elite and the military cannot ignore the ethnic distribution of their society. The decisions fall according to categories: a) those regarded as reliable and therefore trustworthy from the point of view of national security; b) those who are talented and whose their talent can be utilized; and c) those who are regarded as enemies or backward. These categories enable the state elite to make a decision as to whom is to be drafted, who is to be exempted, who can be promoted to higher ranks and who can fill the lower positions in the army. This policy is actually based on what was identified as "an ethnic state security map" which in effect determines the elites' political dependability on ethnic groups. (Enloe, 1980a: 15). Thus a division of labor takes place among the different ethnic groups, each according to this contribution to and weight in the national security (Smooha, 1983-84: 10).

Enloe concludes that the military is "designed by the state elite in such a fashion that it is consonant with that inter-ethnic boundary pattern which makes elites feel most secure. In many cases, the military will also be utilized to create or strengthen that ethnic order deemed most supportive of state security" (Enloe, 1980a: 14).

In surveying the findings of the above two programs of the IDF and other studies on the subject (Azarya & Kimmerling, 1980; Azarya, 1983; Smooha, 1984), one can see that the IDF as an arm of the state elite appears in fact to reflect the policies of the state including its ethnic divisions. The IDF, like the society at large, failed to make the distinction between absorption and integration. To the contrary, given the organizational framework of the Army, the length of service and the intensive experience, the absorption of Oriental Jews has been intensive and deep. It appears that it has been the avowed policy of the state elite in Israel to ensure the deculturization of the Oriental Jews. In effect, the IDF in its educational and cultural programs strived at socializing the Oriental into the culture of the dominant social group. This per se is a positive development since it included indices like male/female equality, democracy, love of the Land, secularism, specialization, family relations etc.. However, this acculturation was accompanied by negative undercurrents of stigmatization and stereotyping which, forty years later, found Oriental Jews more keen than ever to redeem their culture and through it their own self-image and self-respect, especially when everyone realized that the acculturation process was only partially successful. As Enloe put it, "inherent to assimilation processes is the notion that groups are unequal in their cultural claims and status; weaker groups assimilate into norms of the stronger" (1980a: 64).

If the IDF and the state elite were bent on integration instead of absorption and acculturation, the Oriental Jewish heritage should have taken the same place as that of the Ashkenazi heritage in the curriculum of school programs both in the society and the army. The absorptive approach found in the IDF "is fed and feeds the paternalistic approach to Oriental Jews, who are seen as backward people and in need of rehabilitation and thus provides explanations (to justify) their inferior status in it." (Smooha, 1983-84: 15-16).

To that effect, in one of the interviews conducted in the study of 1969, the late Yigal Allon, then Minister of Labor and former Head of the Palmach (a commando or "striking" force of elite military volunteers born in 1941), was asked about the role which Oriental Jews played in the army during the preceeding few years. His answer was that in general the State was very satisfied with their performance. But he did not conceal the fact that one of the reasons for the delayed opening of the war in 1967 was due to the policy makers apprehension of the Oriental Jews' performance in time of war. Then he hastened to add: "Of course, we were very happy to realize that Oriental Jews discharged their responsibilities very well and the victory was theirs as much as it was of all the people of Israel."

Although the picture of the IDF has changed over the years, and Orientals have become an integral and permanent part of the military landscape in Israel and have risen to junior and mid rank officers, nevertheless, Ashkenazi Jews remain over-represented in the higher ranks of the military in general. The only exception to that has been the appointment of the previous Chief of Staff, (1983-87) Moshe Levi, an Oriental of Iraqi origin. Of greater and crucial importance has been the attitude towards the Oriental in the army which has not changed to the degree that one would have expected after years of socialization and acculturation in the country. In 1978, Mordechai Gur, just before ending his tour of duty as Chief of Staff, said of the Orientals:

> When I was Commander of the Gaza Strip I became interested in the mentality of the Arabs. I sat in their classes and conversed with a lot of them. I reached the conclusion that 20-30 years will elapse before the Arab mentality will change. And now the painful side of the issue. The Jews of Oriental origin also will not close the gap for 20-30 years. All the efforts which the people of Israel invested in the Oriental Jews yielded only partial results...Much to my regret, many years will pass before Oriental Jews, including those who receive full education, will succeed in competing with the mentality and the conceptual technology of the West. Here lies the problem. (Haaretz, 21 May 1978).

Smooha finds this statement as further confirmation of the "paternalist-assimilationist concept" which is at the roots of the "state security map." The IDF as an army is built on western conceptual norms, an imperative for the survival of the State. The Ashkenazi community, according to him, provides the "brain and the heart" of the Army whereas the Oriental provides the essential manpower which is needed "to complement the Ashkenazi elite and its needs to be absorbed in it as a condition for its liberation from the Arab mentality and for advancing within the army rank" (Smooha, 1983-84: 17). In addition, since the IDF is an open organization and achievement oriented, "the lower achievement of the Orientals (it is assumed) are due to the slow rate of absorption (lack of adaptation to western mentality)" (Smooha, 1983-84: 17). Such a situation does not hinder the quality of the IDF service since major decisions are made by those in "higher" positions and not by those in the "lower" rungs of the army.

Thus, given the army programs for those Oriental Jews and the structure and length of army service for the general Oriental population, the IDF is found to be even stronger than the civilian society in its absorptive functions. At the same time, due to its stratification, the IDF has not proven itself a promoter of social integration despite the intensive encounter among ethnic groups.

The 1969 study has shown that the acculturation of the Oriental male in the IDF makes him even more predisposed to marry an Ashkenazi woman from the dominant social group. This attitude of the Oriental male has not been matched by a comparable socialization of the Ashkenazi man or woman. As we have seen from the teaching staff of some of these programs, the Ashkenazi woman harbored some stereotypes and her higher education became a source of hindrance for seeking a relationship with an Oriental. On the other hand, the Ashkenazi male has less opportunity to meeting Oriental women due to their low representation in the army as a result of religious exemptions, lack of education, or family traditions. Thus although the intermarriage rate between Oriental and Ashkenazi has reached 22 per cent, it may be assumed that this achievement was only in part due to army service.

Ethnic Stratification

From the specific programs of Camp Marcus School and the David Educational Unit, it is clear that the ethnic stratification in the IDF reflects that of the society at large. As such, these IDF programs are found to perpetuate the ethnic stratification instead of changing it. This was felt also by the recruits themselves.

At the David Unit, the recruits argued that they did not receive the education they needed and which they believed would have given them the necessary minimum mobility during their army service and later in society. Discouraged by this situation, they opted out for menial jobs such as warehouse attendants and drivers for high ranking Ashkenazi officers. Thus the training at David Unit did not break the syndrome of psychological or economic deprivation and it only reinforced their inferiority in the IDF and later on in the society at large. In these cases, the IDF appears to have a clear ethnic stratification. Those from the dominant ethnic group are found in important posts of the decision making process in the IDF and the others are to be found in the lower rungs of the army.

If the conscript succeeds in learning a skill in the army it is usually marketable in civilian society. The level of the skill acquired usually remains the same throughout his army service. The conscript, therefore, does not get moved from one level to a higher one or transferred from one technology to another. In such a framework, the IDF is found to be more rigid than the general society and less open to mobility. Its ethnic stratification would closely resemble a pyramid, at the bottom of which consists of a large number of Oriental Jews with only elementary skills. As one ascends the pyramid one finds less and less of them in the hierarchy of the IDF. This is contrary to the pyramid of today in the civilian society whose shape resembles that of a diamond in which the top and the bottom are smaller than its middle, where most of the population is found.

According to figures cited by Smooha, Oriental Jews in the late 1970s were found to constitute 67 per cent of conscripts and NCO's, 30 per cent of junior officers (who include lieutenants and conscripts), and only 3 percent of the senior officers from the rank of major upward (Smooha, 1983-84: 19). The latter is estimated to be low since 17 per cent of the career officers in the army are of Oriental origin. If this is so the stratification within the IDF is worse than the one found in the society at large. In the case of David Educational Unit and Camp Marcus, the two and three months of intensive training had limited success for the occupational mobility of the conscript and at times even frustrated his expectations.

Conclusions

The case studies and other material obtained from the IDF as well as other studies on the absorption of new immigrants in the IDF do not support the popular image of the role of the IDF in national integration. The successes of the IDF's compensatory programs are found to be at best marginal. Enloe's thesis appears to reflect more realistically the IDF's societal role in Israel. According to Enloe, "conscripts are drawn into an organization that already has a set of symbols, an operational language, an officer corps...conscripts' experience as soldiers becomes one of assimilating, not integrating...their principal contribution is physical manpower." (Enloe, 1980a: 64). Or, as Smooha put it: "The Ashkenazim in Israel constitute a dominant minority whose power is derived from its place as veterans and founders, from its dominant culture and from its political, ideological and class superiority." (Smooha, 1983-84: 27).

The IDF as the people's army and the central and most important institution in Israeli society, seem to reflect the ethnic dominance even more. In order to feel secure, it is only logical that the elite of the army and the state will ensure the dominance in the IDF of that ethnic group which is found dominant in Israeli government which identifies itself with the State and considers itself most talented. Thus, "more than any other institution in the Israeli society, the IDF is led by the sons of Ashkenazi community and functions according to their concepts and norms and in it they exercise the heaviest pressure for the unconditional assimilation of Oriental Jews" (Smooha, 1983-84: 27). As we have seen, the IDF contributed not only to the social and cultural assimilation of the Oriental Jew but also to the "institutionalization of his inferiority class in the army and in the society." According to Smooha, the contradiction between the contribution of Oriental Jews to national security and their status in the society which resulted in the Oriental upheaval and demands for equality and participation may also extend to the IDF especially in light of the "weakening of the IDF immunity in face of criticism" (Smooha, 1983-84: 16).

There are ways of averting the gloomy picture depicted by Enloe and Smooha. That will hinge on the "receptivity disposition" of the Ashkenazi dominant elite to Oriental culture and more rigorous compensatory program of "affirmative action" in the IDF for those Oriental Jews who are in need. The "receptivity disposition" entails the acceptance by the

Ashkenazi of the Oriental culture as part of the political formula. This can be achieved through the inclusion of the Oriental Jewish heritage in all school curriculae, of both civilian and military institutions. Both Ashkenazi and Oriental Jewish descendants would be exposed to the two strands of Judaism, each contributing in its own way to the making of a national Israeli culture and consequently to the promotion of a pluralistic society in Israel. The Ashkenazim obviously played a major role in the establishment of the modern State of Israel. The Oriental Jews can be credited with its expansion, development and supply of much needed manpower that the Ashkenazim at one point were unable to provide. The teaching of Oriental Jewish heritage as part of the national Israeli heritage will gradually eliminate the stereotype of inferiority attached to the Oriental Jew, thereby forming greater equality among ethnic groups.

The other proposal may be more difficult to implement since the IDF prides itself on its meritorcratic values. A suggestion has been made which ironically came from within the ranks of the IDF itself. In 1982, the Deputy Education Officer of the IDF proposed to the IDF to abandon its "minimal model" presently applied in all the IDF education facilities for the absorption of the disadvantaged soldiers and instead use a "maximal model" which in essence is "tantamount to an affirmative action program." Under this maximal model the IDF is required to design an individual program for the special category of disadvantaged Oriental soldier "to achieve the maximum possible in light of his personal potential" (Gal, 1982).

Implicit in this proposal is a modification of the KABA scores, which among other things determine the occupational specialty of the recruit and which have been found to be heavily laden with western values.

If these proposals were to be implemented, these Oriental soldiers who are concentrated in the "non-professional" occupations would be dispersed among other more prestigious occupations of the military. In this way the IDF pyramid would at least resemble more the realities of Israeli society of the 1980 and 90s as demonstrated by Enloe's model. However, both these radical changes towards integration and affirmative action in the IDF will have to wait for the initiative of the dominant state elite.

The IDF, the Intifadah, and the Oriental Jews: A Postscript

Although it is not within the purview of this study to deal with this issue, in light of the role of the IDF in national integration it is important to draw some general lines as to the response of Ashkenazi and Oriental soldiers to the Intifadah during the past two years. No field research has been done on the subject but the few articles which appeared in daily newspapers and the ongoing public debate in the electronic media point in some directions.

In the first place, today Oriental Jews are found along a broad spectrum of politics in Israel, from the far left to the far right. Since the 1973 war, Israeli Jews in general and Oriental Jews in particular have voted increasingly for the right rether than the left. Their vote for the Likud in 1977 showed how the acculturation process has worked. For the first time, first and second generation Orientals born in Israel voted differently from their parents who for the first 29 years of statehood voted for Labor. This pattern of voting was a protest against Labor's absorption policies over three decades and the effects of these policies on Oriental Jews in general. However, in the process of exercising their emancipation, they became identified with Likud's rightist defense and foreign policies. As time passed, it became clear that these policies posed a dilemma for them. On the one hand, Oriental Jews did not abandon easily the culture and traditions practiced in their countries of origin. This is despite Labor's policies of intensive absorption, acculturation and socialization. On the other hand, in their attempt to form an ethnic identity that resembles the dominant social group, Orientals "use the Arabs as a marking-off group in order to bolster their Israeli identity" (Seliktar, 1986: 177). In fact, after the first elections of 1977, Likud could not obtain a clear cut mandate to form a government in the subsequent elections of 1984 and 1988. Instead, Orientals chose to vote for a new ethnic and religious party, Shas (Sephardi

Torah Guardians), which in the 1988 elections more than doubled its strength in Parliament. Orientals had been under enormous pressure to discard their "Arab" identity or to "purge" it before they could legitimately be accepted by the larger society (Roumani, 1988). Yet only 5 to 8 per cent of Orientals are known to live in the occupied territories of the West Bank and Gaza. Among soldiers or reservists one finds Orientals who have been jailed for refusing to serve in the territories.

In the first few months of the Intifadah the IDF had trouble controlling a few "hot heads" who reacted with a strong fist against the stone throwers. Those were usually young Israeli born sabras, Ashkenazi and Orientals who were frustrated and looked for a quick solution to the problem. The IDF had to intervene with an intensive campaign by sending lecturers in the field to "explain" the IDF's policies and to provide a forum for discussion of subjects such as communication and the army, the role of the media in the army, attitudes toward journalists and reporters, etc. In the second year of the Intifadah, no clear differentiation between the two ethnic groups in their response to the events in the territories could be discerned.

It is, however, generally believed that Oriental Jews, especially first generation Israeli born, have reacted stronger than their Ashkenazi brethren precisely because they used the opportunity to bolster their Israeli identity. Two years into the intifadah, there is less of that now. Furthermore, an increasing number of Oriental Jews feel today that the money poured in the West Bank settlements came at the expense of development of the small developing towns where many Oriental Jews are concentrated. It is becoming increasingly clearer, therefore, that there is linkage between territorial compromise or peace and the economic rehabilitation of these towns which are desperately in need of funds.

Recently, the spiritual leader and founder of Shas, the former Sephardi Chief Rabbi of Israel, Rabbi Ovadia Yosef of Iraqi origin, was reported as saying that Halacha, the body of Jewish Law, does not consider territories to be sacred and therefore he does not object to returning them in exchange for peace. In July 1989, both he and his Minister of Interior,

Rabbi Aryeh Deri, met with Mubarak of Egypt; it was reported that both repeated the sentiment. This created a furor in Israel, and a month later on August 13, 1989, in a major rally, Rabbi Ovadia, while not backing away from his original stand, modified it by saying that the return of territories could be done only in exchange for true peace (Jerusalem Post, August 14, 1989).

It is still early to predict whether the Oriental will take the lead in the question of peace. His ethnicity in Israel, his acculturation and absorption in Israeli society, make him ambivalent towards Arabs although not towards Arab culture. However, in recent years there has been an increase in the number of Oriental leaders and Oriental voters in general who have advocated peace and alternative solutions to the Arab-Israeli conflict (Chouraqui 1972; Eliachar, 1975). This may be an indication of a trend in Israeli politics where Oriental Jews will take a lead in foreign and defense policies instead of the back seat as they have done in the past.

References

Azarya, Victor and Baruch Kimmerling. 1980. "New Immigrants in the Israeli Armed Forces". Armed Forces and Society 6: 455-482.

___. 1983. "The Israeli Armed Forces", in Morris Janowitz and S.D. Westbrood, eds., Civic Education in the Military. Beverly Hills, CA: Sage Publications.

Bienen, Henry. 1983. "Armed Forces and National Modernization: Continuing the Debate." Comparative Politics 16, 1(October), 1-16.

Binder, Leonard. 1964. "National Integration and Political Development." American Political Science Review 58: 636-653.

Chouraqui, Andre. 1972. Letter to an Arab Friend. Amherst, MA: University of Massachusetts Press.

Eliachar, Aliyahu. 1975. To Live with Palestinians. Jerusalem: Council of Sephardi Communities.

Enloe, Cynthia H. 1980a. Ethnic Soldiers: State Security in Divided Society. Athens, GA: The University of Georgia Press.

Enloe, Cynthia H. 1980b. Police, Military and Ethnicity: Foundations of State Power. New Brunswick, NJ: Transaction Books.

Gal, Michael. 1982. "Integration of Soldiers from Weak Populations: Summary and Considerations." Ma'arachot 2-14.

Gal, Reuven. 1985. "Commitment and Obedience in the Military: An Israeli Case Study." Armed Forces and Society 11: 553-564.

___. 1986 A Portrait of the Israeli Soldier. Westport, CT: Greenwood Press.

Kimmerling, Baruch. 1979. "Determination of the Boundaries and Frameworks of Conscription: Two Dimensions of Civil-Military Relations in Israel." Studies in Comparative International Development 14: 1.

Lissak, Moshe. 1983-4. "Theses for Discussion or Pre-determined Stands?" Ma'arachot 278: 17-21.

Mosca, Gaetano. 1939. The Ruling Class. Ney York: McGraw-Hill.

Nativ, Moshe. 1983. "Israeli Society in the Reflection of the IDF." Ma'arachot 278: 17-21.

Roumani, Maurice M. 1985. "An Attempt at Promoting Disadvantaged Population in the IDF," in Shlomo Deshen, ed., Half the Nation: Studies in the Culture and Status of Middle Eastern Jews in Israel. Ramat Gan: Bar-Ilan University.

___. 1988. "The Sephardi Factor in Israeli Politics." The Middle East Journal 42: 423-35.

Salomon. Nissim. 1982. "Education in the IDF: Fixed Trends and Processes of Change." Ma'arachot 283.

Seliktar, Ofira. 1986. New Zionism and the Foreign Policy System in Israel. Carbondale, IL: Southern Illinois University Press.

Shor, Yitzhaq. 1982. "One Quarter of the IDF Recruits have not completed schooling." 'Al-Hamishmar.

Smooha, Sammy. 1983-4. "Ethnicity and the Military in Israel: Theses for Discussion and Research." State. Government and International Relations.

Zamir, Israel. 1984. "The IDF attacks the Educational Gap". 'Al-Hamishmar.

4

GREECE AND TURKEY: THE MILITARY AND SOCIAL INTEGRATION

James Brown

Greece and Turkey hover on the threshold between Europe and the non-Western world. Geographically, Greece is European, whereas Turkey straddles both the European and Asian continents. Both of these nations possess all the political institutions of modern democratic states, but their institutions have not operated in the fashion of their Western European counterparts. Even more disturbing to Western observers have been the periodic suspensions of representative institutions by both the Greek and Turkish armed forces.

The events of the World Wars and the complexity of the international scene, the growing urbanization of both Greek and Turkish societies, and the appearance of a large mobilized population have produced groupings that cannot or do not want to be accommodated in the traditional parliamentary system. These occurrences have given rise in both countries to organizations at the extraparliamentary level which are characterized by hierarchic authority and ideological coherence. These extraparliamentary groups have attempted at times to undermine the traditional governing institutions, thereby pitting themselves against the military who view themselves as the protectors of the state.

The impetus to unite and create a modern state, in both Greece and Turkey, came essentially from the outside. Both nations are remarkably homogeneous with a single language

81

and culture, a single religion, and a common revolutionary past. However, such parallels are misleading. Political turmoil, schisms in the polity, and military intervention have been persistent in each nation. The military, but most especially the officer corps, have played both direct and indirect roles in the political processes. However, in both cases the armed forces have also played a positive role in generating national values, undertaking philanthropic projects, and serving as mechanisms of political socialization.

The focus of this paper, therefore, is to examine both the Greek and Turkish armed forces on their roles as actors in the political system and, more specifically, as forces of social integration. This process involves the mass population and presumes meeting their demands for rewards, recognition and status, at least to some degree. Jean Bodin observed:

If the lower classes are once armed and not constantly employed against the enemy, there is no doubt that sooner or later they will try, and succeed in, changing the form of government to secure a share for themselves...If only the ruling class is armed, one day it will be defeated in the field... (McWilliams, 1967: 22).

This statement necessitates an analysis of the recruitment patterns of the military and an examination of the representation of all societal groups (ethnic, linguistic, tribal, geographic) in the armed forces. Is there a binding national ideology that transcends that of nationalism? Lastly, what are the politico-military implications of these findings in holding and retaining power for Greece and Turkey?

Greece

In its 160 years of independence, modern Greece has been a scene of turmoil and political instability. Foreign interventions, coups d'etat, and war have been common in its history.

On the surface Greece appears to be homogeneous in culture, language and religion. Ninety-eight percent of the population speaks Greek and practices the Greek Orthodox faith. The remaining two percent of the population consists of

Turks, who are practicing Muslims (1.3%), and some smaller ethnic groups of Slavs, Vlachs and Albanians. None of these groups poses any threat to Greek national unity or security. However, these groups are affected when issues arise between Greece and one of her neighbors (e.g. Greek-Turkish relations and Cyprus). This homogeneity is misleading and belies a nation of dissension lacking a national consciousness and mutually compatible values (Legg, 1969: 82). Historically, there were cleavages in Greek society along the lines of language and culture, and even religion has not been the unifying force that it is assumed to be. Presently, there is also a growing tension between urban and rural sectors of Greek society; and "even within these groups there is little agreement in the meaning of 'Greece' or the nature of the regime" (Legg, 1969: 82).

Given this disagreement and the underlying norms and goals of the Greek nation, it is not surprising to find conflict over a political formula of governing Greece. A closer examination of Greek political history will reveal that Greek politics are feudal in character and that relationships between politicians, voters, and bureaucrats are based, even today, on clientage and patronage ties. There seems to be little consensus "on the meaning of Greece or the nature of the regime" (Legg, 1969: 82-83). On the one hand, there are the Western institutions (i.e. parliament) while on the other hand, there is the traditional Greek pattern, with its clientage networks that extend from the village or city to parliament and the bureaucracy and require the satisfaction of individual demands. As Legg described the process, "These personal demands are directed at governmental output; they arise because of the basic distrust present in Greek society, because of the malfunctioning of the state machinery, or because of formalistic legislation" (Legg, 1969: 184).

A Historical Perspective

Greek society in the nineteenth century contained a distinct but small civilian elite whose power lay in its wealth, education, mobility and foreign alignments.

A predominantly rural society, one fragmented into kinship

groups, integrated only through competitive institutions, and linked with other groups by patronage...(which) affected the formation of a regular army profoundly (Veremis, 1978).

Political and military leadership tended to coincide with very little differentiation between military and civilian elites. Until 1897, officers tended to act in unison with politicians to enforce constitutional reform or even to produce changes in government. The penetration of the armed forces into the political arena was ingrained in the officer corps and set the tone for the twentieth century. Historically, the political allegiance of the military has been dictated by alliances with the monarchy, politicians, and political parties.

Beginning with the 1909 intervention, the Greek officer corps has been, more or less, the focal point for securing and maintaining political power in Greece. From 1912 on, the armed forces were split into two rival groups: the Republicans and the Royalists. Veremis (1978) describes this period in these terms:

> The proximity of the military to civilian mentality may be illustrated by the fact that during the interwar period, coups d'etat had almost acquired an informal status as a means of political pressure. Since the social fabric was never seriously threatened by the coups, civilian society had more or less adapted itself to such practices as a way of promoting private interests...Politicians were, by 1935, conditioned to the idea of military intervention and qualified coups (532).

The Nazi occupation of Greece, and the resulting fragmentation of the armed forces during the war, was a break in history but not in tradition. Shortly after the liberation of Greece from German control in December 1944, bitter fighting erupted in Athens between Greek communist elements, and Greek national forces backed by the British Army. The costly civil war ended in 1949, but in the interim, King George returned to the throne, the communists lost the struggle, and, after the proclamation of the Truman Doctrine (March 1947), the United States became the primary supplier of military equipment and professional training to the Greek armed forces.

The Civil War left an indelible impression on the officer

corps and marked a new era for them, a period in which the armed forces emerged as a major power bloc, independent of traditional parliamentary factions and leaders. The military was now viewed as a symbol of national unity and guardians of the state, who saved the nation from a communist takeover. These views were permanently etched in the officer corps. They showed considerable social and educational homogeneity and were politically conservative. Other than this conservative, anti-communist bent, the Greek military did not develop nor possess an ideology comparable to Kemalism, which is today an indigenous asset for Turkish society, its armed forces and ruling elite.

The April 1967 Coup

The conditions that brought about the April 21, 1967 coup d'etat were quite varied. Six reasons appear discernible: the perceived communist threat, the political incompetence of parliamentarians, the decline in the growth rate of the economy, the social and moral decay of Greek society, Greece's geostrategic role in NATO, and the professional grievances (promotions and salaries) of certain elements of the officer corps. The 1967 coup was more acceptable because most of the Greek population recognized the politico-socioeconomic factors that led to this action. The longest period of military rule in Greece (1967-1974) culminated when the Junta 2 launched an ill-fated coup against Archbishop Makarios and the Cypriot government in the hopes of uniting Cyprus with Greece (ENOSIS). The plot failed and the Greek armed forces were totally disgraced and forced to return to the barracks.

The new civilian government, led by Konstantine Karamanlis, inherited a demoralized military. Using the "carrot and stick" approach, the government proceeded to purge unrepenting officers while also attempting to establish a cordon sanitaire about the military, thereby neutralizing them. The election of 1981, which brought Andreas Papandreou to power once again, created consternation within the ranks of the officer corps. In an effort to allay any concerns on the part of the military, Papandreou resurrected two foreign policy issues that strongly appealed to their nationalism. First, and foremost, was the state of Greco-Turkish relations. Since 1974, all Greek governments have developed a consensus concerning the "Turkish threat". This preoccupation with Turkey is over several key issues: the Cyprus imbroglio that was precipitated in 1974 by Athens, the ongoing dispute over air, sea, and exploration rights in the Aegean Sea. Papandreou dismissed any threat from the communist north and saw the east (Turkey) as Greece's major threat.

Greece's dealings with NATO and the United States was the second issue of concern. Papandreou followed a very calculated but contradictory policy. On the one hand, he successfully negotiated with the United States in 1983 a 5-year Defense and Economic Cooperation Agreement (DECA), while at the same time being very supportive of the Soviet Union in the foreign policy area (e.g. Poland, nuclear free zone in the Balkans). Greece and the United States underwent a seventeen-month period of discussions over the four major bases and twenty smaller installations. These are a large part of the American presence in the eastern Mediterranean. The Greek side demanded that the U.S. presence be removed from Ellinikon Airbase and that security guarantees be given to Greece against perceived threats from Turkey.

Although Papandreou attempted to neutralize the armed forces, in reality the Greek military today, and in particular the officer corps, is as easily politicized as in earlier periods of Greek history. Promotions and officer retirements, especially at the senior ranks (Lt. Colonel and above) are dependent upon an officer's support of Papandreou and PASOK or the appearance of complete neutrality. For one to hold power presently in the armed forces, a certain political coloration is required. Purges and massive dismissals at the general's rank have cowed the officer corps into submission.[3]

Indeed, in the post World War II era, the Greek armed forces believe that they are the embodiment of national ideals and that they are a fixed and integral part of Greek history. The armed forces, modern Greece, and her history have become inseparable in the eyes of the armed forces. It is commonplace for officers to boast that they more clearly represent the mainstream of Greek values and the views of a majority of Greeks than do the elected politicians and bureaucrats. This poses major difficulties for civilian governments in maintaining supremacy over the military establishment. An examination of the socioeconomic background of the officer corps will reveal that the military and civilian elites are not integrated and view each other with great suspicion and distrust, thereby rendering rule by parliamentary government difficult.

A Profile

The service academies are the basic sources of recruitment for Greek officers in all three branches of the military. They are the institutions that recruit and socialize the military elites. They view officership as a "sacred mission" and the military cadets are called *evelpis*, meaning "the best hope for the nation." At times, governments have used the academies to indoctrinate cadets in the political values of an existing regime rather than in professional ethics; for example, during the late sixties and early seventies a very pro-Junta view and a denigration of all elected officials was commonplace.

The military recruits cadets primarily from agricultural and small towns, except for the Navy, which tends to recruit from urban areas. Close examination of the data in Table I shows that the Army and the Air Force tend to recruit from the same general areas in Greece, but the Navy draws from central Greece and Euboea, the Aegean Islands, and the Athens area. A partial explanation for the over-representation of Naval cadets from these areas may be that the recruitment patterns differ along socioeconomic lines. Historically the Navy has been viewed as the more prestigious service, with links to royalty, shipping magnates, and the more affluent sectors of Greek society.[4] This situation still prevails to a lesser degree.

The Army and the Air Force tend to draw from relatively poor economic regions whose inhabitants are attracted by the mobility, security, and prestige offered by the military. For example, the regions of Macedonia, Thrace, Epirus and Crete contribute a high proportion of Army and Air Force cadets. In addition, these regions, except Crete, border on countries long

Brown

TABLE I

GEOGRAPHIC ORIGINS OF MILITARY ACADEMY CADETS CLASS OF 1985

	Army	General Navy	Air Force	Population
Peloponnesos	15.26%	11.47%	17.01%	10.25%
Central Greece and Euboea	20.82	30.96	23.50	11.39
Macedonia	20.24	5.73	16.66	21.55
Thessaly	12.11	1.63	14.73	7.54
Thrace	6.80	2.45	4.83	3.77
Epirus	4.81	0.82	2.63	3.55
Crete	8.46	4.09	6.84	5.21
Ionian Islands	1.35	1.16	1.20	2.09
Aegean Islands	2.80	4.09	1.45	4.76
Athens Metropolitan Area	7.30	37.70	11.05	29.96

Source: Meleti ths Sxoles ton Evelpidon (Study of the Army Academy) a yearly study, Greek General Staff, Research Division, Athens, Greece.

hostile to Greece. Thus war, with its attendant chaos and devastation, has become a way of life for the people of these regions, all of which are known for their "heroic" values.

Specifically, the Army and Air Force cadets are recruited from rural areas and one may infer that they tend to come from humble circumstances where the fathers' occupations may be those of farmers, tradesmen, or public employees. Officers' careers serve as convenient vehicles for social mobility otherwise unavailable. There are two principal reasons why the military has become an avenue of social mobility for young men from rural areas. First, economic and technological developments and the expansion of civilian bureaucracies have broadened the opportunities in nonmilitary careers for the best educated, who tend to reside in the urban areas. Second, public schools have given the rural lower classes the opportunity to receive the academic training needed to qualify for the academies, thereby enhancing their future career goals.

Another element indicating social background is the proportion of cadets and officers who enter through self-recruitment as sons of professional officers. The percentage is highest in the Navy (16.7%) and considerably less in the Army (7.7%) and the Air Force (5.7%). The Air Force's relatively small percentage may reflect its being a fairly new service that has not had time to develop self-recruitment patterns.

The rural social backgrounds of the officer corps, especially the Army, coupled with its lower class occupation origins, support the contention of a "fundamentalist orientation and lack of integration with other elites, especially political elites" (Janowitz, 1977: 134).

As indicated elsewhere, kinship ties and family connection characterize modern Greek politics and life. This also extends into the military. No matter the complexion of the regime, political leaders have cultivated clientage within the military, and officers may have resented the system; however, they have not hesitated to use the exchange of favors (rousfeti) as a means of career advancement.

For most Greeks the basis of any regime and even obedience to authority depend on personalities and clientage networks that tie individuals to particular incumbents. In this kind of setting, actual inheritance of political position is common.

The significant number of parliamentarians from families with histories of political involvement indicates the existence of a network of political influence at provisional and local levels (Legg, 1969: 249-316). In fact, during the 1946-1965 period Legg found that political families have retained strong holds on political offices (Legg, 1969: 307). The advantages offered by particular regional backgrounds are related to clientage, and most Greek politicians retain strong ties with their place of birth, even if they live elsewhere. For example, Karamanlis' most recent government (1977) had seven cabinet members from Macedonia; only Athens with ten posts had more.

One major note of importance is that politicians and political families tend to represent the more affluent sectors in regions, while members of the officer corps are recruited from lower socioeconomic stratas.[5] These differences color the officers' perceptions of each other and the various Greek

institutions, resulting in political mischief on the part of the
armed forces.

The Military: A Socializing Experience

All Greek citizens capable of bearing arms are required by
the 1975 Constitution to contribute to the defense of the
country.   Conscripts comprise some 66 percent of the armed
forces, serving tours of duty which vary depending upon the
service (Army 21 months; Navy, 25 months; Air Force, 23
months).   Universal conscription maintains the armed forces.
However, in the next decade meeting manpower demands will
be quite difficult for Greece because of zero growth in the
population.   Presently, Greece is experimenting by permitting
women to join the armed forces.   It is hoped that enough
women will enter to meet future manpower demands.

The conscription system operates effectively and is an
accepted part of Greek life, particularly in rural areas where
military duty may provide a conscript with his first venture
away from home and with training opportunities not otherwise
available.   Men from the rugged mountain districts of northern
Greece generally expect Army service, and those from coastal
areas and islands that are familiar with the sea and ships are
more likely to receive Navy duty.

By the 1960s, universal military training was credited with
having virtually eliminated illiteracy among men under the age
of forty.   Recruits deficient in basic literacy skills receive
instructions in reading and writing as part of their military
training.   The percentage of recruits that fall in this category is
less than 3 percent.[6]   Personnel also receive technical and
administrative training which contributes skilled workers to the
economy.   In 1985, a law was introduced that permits
conscripts who are attending educational institutions to
complete their basic military obligation by attending two month
training sessions during the summer.

Since the early 1980s, military units are fully integrated
without distinction being made based on one's political
leanings.   This has not always been true in the Greek armed

forces. During and immediately following the Civil War, draftees and left-wing officers were separated from the others and sent to a small island off the tip of Attica for military training and indoctrination. Upon completion of this training, draftees were put into special units that were widely known for their severity, and they were usually assigned to the most unpleasant duties in the northern mountainous areas of Greece. Up until the Papandreou government came to power, it was a fundamental and firmly adhered to assumption of the officer corps that no person with Leftist political ties could possibly be loyal to Greece. In 1983, general amnesty was granted to those Greeks who had fought in the Civil War on the side of the Communists. The assumption that conscripts and officers had to have a clean police record, meaning that he had no Leftist leanings, in order to be eligible to enter the military, was cancelled. Furthermore, the social science curriculum at the academies has been changed. The historical discussions of the Civil War have been tempered and the strong anti-communist rhetoric that was part of the curriculum in the post World War II period has been greatly softened. These changes have created some consternation with many senior officers (Lt. Colonel and above), most especially those who either fought against the EAM/ELAS elements or were some way affected by the Civil War.

The armed forces also stress their economic and social missions. Through the Military Reconstruction Projects Service, military personnel take part in public works, road building, housing construction and land reclamation, not to mention assisting civilians during instances of natural disaster. Today, in spite of seven years of Junta rule, the nation has a very positive view of its armed forces, who are always an integral part of all celebrations on patriotic days. The embodiment of the national ideals of the Greek military and their link to Greece's heritage and War of Independence in the nineteenth century are the Evzones (Honor Guard), who wear the traditional costumes of the mountain warriors and are prominently seen guarding the Tomb of the Unknown Soldier in Athens.

The preoccupation of the Papandreou government with the perceived "Turkish threat" and the varied social and economic functions of the Greek armed forces, as described above, has furthered Greek national self-esteem and in turn has linked the Greek nation more closely to its armed forces. A military that was totally traumatized in 1975 has now become, once again, the embodiment of Greek nationalism and the "barracks" appear to be a far safer haven for them than excursions into the political arena.

Turkey

The Republic of Turkey (Turkiye Cumhuriyeti) was established on October 29, 1923 under the control and leadership of Kemal Ataturk. The creation of this new republic was the result, as Bernard Lewis expressed it, of "a victory of Turks over Ottomans." Ataturk's avowed goal was to create from the shattered remnants of the Ottoman Empire a new country and society patterned directly on the countries and societies of Western Europe. The driving force of these reforms rested in Ataturk's "Six Arrows" - secularism, republicanism, statism, populism, nationalism, and reformism. Of these, secularism was, and remains today, the most significant and perhaps the most controversial aspect of Kemalism. This meant a separation of all Ottoman tradition from the state, which was to take a European form. Ataturk was not hostile to cultural and religious tradition, but he considered them to be obstacles to national survival. Therefore, Ataturk and his colleagues may be viewed as extending the thought, behavior, and values of the secular elements of the Republic. To this date the bureaucratic and military elites and most political party leaders embrace this political and social philosophy.

What appears to be an officially mandated homogeneity is deluding; domestic discord and violence have been vexing. This has been especially true in the post World War II era, and in particular during the 1970s, when Turkish society was fractured. In short, the domestic situation by the late 1970s reached a critical stage. Against the backdrop of ten years of coalition governments, political polarization (which included fractious sectarian, linguistic, and religious rivalries), urban terrorism, and fundamental socioeconomic problems, there rose the specter of anarchy or military rule.

On September 12, 1980, the Turkish armed forces once again seized power to save the nation from severe social, economic, and political chaos. Previously, in 1960 and in 1971 the military had intervened because of the inability and ineptness of Turkey's politicians and political parties in maintaining a viable political system.

Historically, the Turkish military, going back to the beginnings of the Republic, adhered to the principles imparted by Ataturk that the armed forces were to stay out of politics. But a closer reading of Ataturk would suggest otherwise. In a speech that he delivered in Konya he states that:

> This apolitical army, totally subordinate to the civil power, is at the same time entrusted with the mission of securing the unconditional defense of the political institutions of the state against both external and internal attack (Clapham and Philip, 1985: 49).

Thus the armed forces have reserved the right to intervene should developments threaten Ataturk's policies and goals, thereby threatening the Turkish model of democracy.

In undertaking the 1980 coup, General Kenan Evren, then Chief of the General Staff,[7] pointed out, "We have not eliminated democracy. I would particularly like to point out that we were forced to launch this operation in order to restore democracy with all its principles, to replace a malfunctioning democracy" (FBIS, 1980: T-1).[8] Within a matter of three years, elections were held. Turgut Ozal and the Motherland Party took over the reigns of government; civilian rule was again restored. There is no doubt that President Evren and the Turkish armed forces are committed to perpetuating Kemalism, but, most especially, the principles of secularism, republicanism, and nationalism.

Socio-Cultural Cleavages

Nearly all members of Turkish society are professing Muslims with about 93 percent of the population proclaiming

Turkish as their mother tongue.[9]   The question "Who is a Turk?" nonetheless evokes answers that suggest that the society is a pluralistic mosaic of diverse and at times contending groups.   Numerous observers, both Turkish and foreign, note that an overwhelming majority of the population -- estimated in 1987 at about 52 million -- accept as "true Turks" only those individuals whose mother tongue is Turkish and who adhere to Sunni Islam.   Nevertheless, an estimated 6 to 8 million Kurds (dogulu), who primarily reside in Southeastern Anatolia and are loyal to Ankara, have over the years, to some extent, resisted assimilation, because of their traditional orientation and tribal communal structure. This is the largest minority group in Turkey, consisting of about eight percent of the population.   The overwhelming majority of these peoples claim Kurdish as their mother tongue (Nyrop, 1980:69).[10] Most are Sunni Muslims, but an estimated three million are members of a deviant sect of Shiite Islam, known as Alevis.   These Kurdish speaking Alevis are thus twice removed, linguistically and religiously, from the Turkish speaking Sunni majority.   In addition, some one million or so of Turkey's Arab population (about three million) are also Alevi.   They reside in or near Hatay Province, which is an extension of the Syrian plain, and most likely maintain ties with the Alawites of neighboring Syria.   Beyond these two groups, there is also a small scattering of Greeks, Jews, Laz, and Armenians.   Other than the Kurds, no other group could possibly pose a threat to Turkish national unity.

A Profile of the Officer Corps

To what extent are these groupings represented in the Turkish armed forces?   A starting point is to examine the social origins of the Turkish officer corps which "is a powerful key to understanding its political logic" (Janowitz, 1977:81).[11] Officership in Turkey is viewed quite positively and is considered a very respected profession.   The service academies (known formally as the War Colleges) are the primary sources of this recruitment and they are, furthermore, the institutions that socialize the military elites.   These academies not only instill professional ethics, but throughout the four years that a cadet is in attendance, he will be inculcated with the values of Kemalism and will study in detail the history of modern Turkey. This is a required part of the curriculum and continues throughout an officer's career.

We might believe that the Turkish officer corps is heavily recruited from rural areas, as was the case in Greece. Although most of the Turkish population presently resides in the rural areas, data suggest that cadets and the officer corps' origins are from the urban areas. In fact, the five major cities in Turkey (Istanbul, Ankara, Izmir, Adana, Konya), which contain sixteen percent of the total population, contribute the lion's share of cadets for the Army and Air Force.

Table II examines whether or not there are any geographic concentrations of cadets and the officer corps. It is clear that the Army cadets favor Central Anatolia, while at the same time, this region represents 29.2 percent of the total population. This region is a relatively poor economic area, and the mobility, security and prestige offered by the military appeal to the inhabitants. A stronger argument can be made that Central Anatolia is closely associated with Ataturk and the Revolution, a revolution which even today engenders great appeal for the general population. The region of the Aegean/Marmara Seas is greatly skewed in favor of the Air Force and Navy cadets, with 73.7 percent and 56.5 percent respectively, whereas only 29.1 percent of the total population of Turkey resides in this region. This area contains two of the most cosmopolitan cities in Turkey -- Istanbul and Izmir -- which are less "tradition bound" than other cities and areas of Turkey. This finding further substantiates our earlier data which proposes that the major cities contributed substantially to cadet recruitment. In addition, both the Air Force and Naval War Colleges, plus numerous major installations of both services, are located in these two cities, which fact adds further weight to our argument. Lastly, we can proffer that this entire region has a seafaring tradition, which perhaps facilitates the Navy's recruitment process. It is also evident from Table II that both Eastern Anatolia and Southeastern Anatolia regions are under-represented in all recruitment patterns. This is explained by several factors: the geographic remoteness of the

TABLE II

GEOGRAPHIC ORIGINS OF MILITARY ACADEMY CADETS 1982-1984

Region	Army	Navy	Air Force	Total Population
Central Anatolia	44.5%	24.2%	13.4%	29.2%
Aegean/Marmara Seas	24.6	56.5	73.7	29.1
Black Sea	12.1	8.5	5.0	12.7
Mediterranean Sea	9.2	4.0	4.1	12.4
Eastern Anatolia	7.6	5.8	2.8	10.4
Southeastern Anatolia	1.8	1.0	1.0	6.3
Numbers	(951)	(718)	(716)	40,347,719

Source: Turkiye Istatistik Yilligi 1983 (Statistical Yearbook of Turkey, 1983) State Institute of Statistics, Ankara, Turkey, 1983.

areas, economic underdevelopment, and a higher rate of illiteracy than the rest of Turkey. Southeastern Anatolia is populated largely by Kurdish speaking Turks who, over the years, have to some extent resisted assimilation, due to their traditional orientation and tribal communal structure.[12]

A part of Southeast Anatolia, consisting of the provinces of Gaziantep, Hatay, Mardin, and Urfa, is home for the bulk of Turkey's Arab speaking minority. Although this minority possesses a language difference -- an ethnic distinctiveness fed by cultural and economic differences -- the recruitment patterns for Army cadets are not affected, while those of the Air Force are skewed negatively.

Are there any linkages in recruitment from the regions where Ataturk initiated the Revolution? The nationalist movement was launched at Samsun on the Black Sea in May of 1919. Subsequently, nationalist congresses were held in the following months in Erzurum and Sivas. It is also from the Eastern Anatolian provinces that Ataturk raised a cohesive army which would go on to defeat the Greeks and ultimately establish the first Turkish Republic. Our data reveal that no significant biases in recruitment of cadets or officers have emerged from these regions. In fact, the Army, with its close proximity to these historical events, is somewhat under-represented in comparison to the geographic regions.[13]

Most of the cadets from the academies are recruited from the urban areas and have lower-middle to middle class origins (50,000-200,000 Turkish Lira monthly). Cadets from the upper strata of Turkish society do not find a military career rewarding. There are, however, 37.7 percent of the Army cadets and 28.9 percent of the Air Force cadets who are from the lower socio-economic strata, suggesting that a career in the military is viewed as an avenue for social mobility and prestige, and that social background will not hinder success in their careers.

Another dimension of social origin is the view of the occupation of the cadet's father as a measure of social origin. It may also be employed as an indicator for locating an officer's parents at some point on the social pyramid. Accordingly, the data in Table III indicates that military, gendarmerie, and civil servants produce the largest percentages of all officers in the Turkish armed forces, approximately some 40 percent or more for each service. Thus a major aspect of social origin is through self-recruitment, that is, "sons of military" and civil servants.

Both of these elites hold a significant place in the history of modern Turkey. They imbue themselves with guardianship responsibilities to the political system and the state, even though they represent a minority of the population. Both have been the primary agents contributing to the political stability of Turkey. If counter elites enter the political system, as in the 1960s and 1970s, and appear to stray afar from Ataturk ideology, this conceivably draws them into the political arena. Thus the "guardians of the flame" of Kemalism assure through self-recruitment the perpetuation of the Revolution and imply a direct stake in the existing order. This conclusion is not at all consistent with findings by other scholars regarding elites and military intervention (Janowitz, 1977a: 50).

Additionally, farmers' and laborers' sons use this career path as a springboard to upward mobility. Conversely, it is not at all surprising to find that professionals as a class do not view a military career positively. In general, these background characteristics imply that the officer corps has a stake in the existing social and political order, and Kemalism is the guiding

TABLE III

OCCUPATIONS OF CADET OFFICERS' FATHERS 1982-1984

	Army	Navy	Air Force*	Total Population
Military and Gendarmerie **	20.5	26.1%	19.9%	1.8***
Civil Servants	17.9	29.4	19.9	0.9
Laborers	13.4	16.8	11.1	21.2
Professionals	7.3	3.9	5.5	5.0
Tradesmen/Merchants	11.0	8.7	13.8	12.2
Farmers	8.4	–	9.2	57.2
Others	21.3	15.5	20.3	3.7
Totals	(3123)	(718)	(716)	19,212,183

*Includes the cadet classes from 1980 through 1984.

**This category includes not only officers, but non-commissioned officers, as well as retirees from both groups.

***There are approximately 78,500 officers and non-commissioned officers in the Turkish armed forces. This percentage figure is not calculated as part of the other occupations.

Source: Turkiye Istatistik Cep Yilligi, 1984 (Statistical Pocket Book of Turkey, 1984). State Institute of Statistics, Ankara, Turkey, pp. 30-31.

philosophy for this involvement. It is clear from the data that the recruitment of the officer corps clearly is lower-middle to middle class, broadly recruited throughout Turkey (except South eastern Anatolia), suggesting a perpetuation of kinship and personalties to Ataturk and the Revolution he wrought. Both of these groups, particularly the military, tend to view political parties and politicians as having narrow, selfish and personal interests that have not in the past been in the best interests of Turkey.[14] There is also a smaller group who became officers for the upward mobility offered by the military. However, the glue that binds these officers to the previous groups and to other societal elites is Kemalism, the philosophy that is inculcated in cadets and officers throughout their careers.

There has been no precise method of determining when restraint yields to guardianship, and decisions to invoke the latter responsibility have been painfully made. At the least, such justification would seem to require a lengthy period of failure to deal with urgent problems on the part of the normal political process. If the past is our guide, such a military move would come only after warnings and open indications of military displeasure. Basically, the officer corps, which is generally representative of Turkish society and is held in high esteem by them, is sincere in its attachments to the democratic process and its concern that it work effectively. Indeed, the corps, like the remainder of the Turkish population, appears to accept the principles of multi-party politics and an elected parliamentary assembly.

A Socializing Experience

According to Article 73 of the Turkish Constitution, "National service is the right and duty of every Turk. The manner in which this service shall be performed, or considered as performed, either in the Armed Forces or in public service shall be regulated by law." Under the implementing legislation, or Military Service Law, all males registered are technically liable for call-up to perform military service. The service obligation is for eighteen months for all services beginning at age twenty, of which three months is in basic training in a highly disciplined, structured, fifteen-hour-a-day environment. There are no minimum educational or intelligence requirements for military duty. During the three months of basic training, and as long as he continues in the military, the conscript is inculcated with a sense of Turkish history and a fundamental understanding of the values and principles of Kemalism. As we have noted earlier, the officer corps is also steeped in Kemalism on a continuing basis. This indoctrination not only provides the enlisted personnel with an understanding of the governing principles of modern Turkey, but further assists in supporting and linking the officer corps to the rank and files in a common ideology. Once the conscript is discharged from the service, he carries the values of Kemalism into Turkish society and life.

Of the total number recruited annually, some ten percent

are illiterate. These persons before they leave the service will be provided with a fundamental education. Additionally, 120,000 trainees leave the services annually. Of these, most have learned a particular skill that they can utilize to become productive and contributing members of society. Based on the Internal Service Regulations, the Turkish armed forces are required to participate in civic action programs, most especially in times of natural disaster (e.g., earthquakes, fires, floods, etc.) Since 1984, surplus numbers of conscripts have been employed in national service projects such as road maintenance, construction, mining, logging and health programs. Also, special military exercises take place that bring the medical corps to remote regions of the country. During such programs, health care (including surgical procedures) is provided to the inhabitants. Lastly, most of the demands for blood by the Red Crescent are met by the armed services.

The Turkish military are utilized for ceremonial purposes and are always an integral part of the formalities on patriotic days. There is no doubt that these activities enhance nationalistic feelings and thereby contribute to the popular sense of self-esteem.

Although the Turkish government provides no data on ethnic, religious, and linguistic minorities, it is safe to surmise that national service is viewed very positively as an essential part of one's citizenship. Few individuals fail to fulfill this civic responsibility. Those that do are most likely to escape abroad or are already living abroad and have no intention of returning to Turkey. In fact, many sons of Turkish nationals, who are living abroad or who have emigrated, return to Turkey to complete this service obligation and upon completion view this accomplishment with great pride.

Conclusions

Both Greece and Turkey appear on the surface to be homogeneous in culture, language, and religion. However, in each country there exist undercurrents for political turmoil and schisms. Military interventions are part of the political landscape.

In the case of Greece, the military, historically and today, is a major power bloc in holding and retaining power. They view themselves as symbols of national unity, guardians of the state and saviors of Greece from Communism. The officer corps is recruited from the lower socio-economic stratas of society, and they view this service as a vehicle for social mobility. On the other hand, the political and bureaucratic elites tend to represent the more affluent elements of society. These differences color the officer corps' perceptions of each other and the various Greek institutions, and at times have placed the civilian and military institutions on a collision course. However, the obvious question of whether the military will, in the future, seek to intervene in politics is an open question. It can only be answered by the degree of structural change that takes place in Greek society which will render the officer corps immune from extra-parliamentary arbitration.

In sharp contrast, Turkey's polity possesses two outstanding characteristics. First is the presence of a strong personality, Kemal Ataturk, who shaped, defined, and established principles of behavior for the governing of Turkey and the armed forces. Second, the civilian and military elites' socio-economic backgrounds are similar, and they both hold a significant place in the history of modern Turkey and subscribe to Kemalism. They imbue themselves with guardianship responsibilities and contribute to the stability of the state. It is only at last resort, and reluctantly, that the armed forces intervene if counter elites enter the political process and stray from the tenets of Kemalism.

On the other hand, Turkey does possess a fairly large minority of Kurds who have not been willing to assimilate to Turkish life, society and politics. Armed clashes between Turkish security forces and Kurds persist today in Southeastern Anatolia. This minority group is a possible threat to Turkey's national unity. Greece, in sharp contrast, lacks such a cleavaged group. The Turkish minority in Greece might prove a slight annoyance, but nothing to the extent of some of the Kurdish groups in Turkey that have engendered separatist attitudes. These elements are apparently benefiting from substantial foreign training and material support abetted by cross-border safe havens in western Iran, Syria and northern Iraq.

It does appear that both the Greek and Turkish armed forces serve as vehicles in social integration of the masses. Political education and socialization are positive instruments, utilized by both armed forces in fostering the embodiment of national ideals in both the conscripts and officer corps. Furthermore, universal military training is credited in both Greece and Turkey as having assisted in eliminating illiteracy. In addition, conscripts receive technical and administrative training while in the service, which not only contributes positively to the economies of both countries but further provides positive links to their national institutions.

Finally, there is no doubt that all of these activities provide acceptance and help nurture unity, self-esteem, and nationalism in Greece and Turkey's mass populations, thereby creating a latticework for social integration.

Notes

1.   Even today conflict exists between speaking the purified language (katharevousa) and the common language (dimotiki).

2.   The primary instigator of this coup was Brigadier General Dimitrios Ioannides, head of the military police (ESA).

3.   Beginning in February 1984 and subsequently in December 1984 and March 1985, a total of 6 Lt. Generals, 37 Major Generals and 79 Brigadier Generals were retired.   Also, the Chief of the General Staff (General Nikolaos Kouris) is an Air Force officer, whereas historically this billet belonged to the Army.   All the major slots of the armed forces are headed by Papandreou supporters or political neutrals. Professionalism is not a major criteria for selection to these posts.

4.   The Navy was highly implicated in the king's counter-coup of December 1967.   During the entire period of the Junta, the Navy was, for the most part, suspect.

5.   Cabinet ministers and their families tend to favor certain educational institutions (Protypon Gymnasium and

Athens University). In addition, they travel and study extensively abroad. Furthermore, they belong to voluntary and elite organizations (the Athenian League, Society of Athenians, etc.). However, intraelite conflicts and antagonisms exist between old aristocratic families (tzakia) and the newly mobilized strata of entrepreneurs (neoplouti). These elites are not homogeneous in outlook or philosophy, but they have interlocking relationships. The cabinet ministers tend to be all males, university educated and predominantly lawyers.

6. Telephone interview of Colonel Mathew Halkiadakis, Military Attache, Embassy of Greece, Washington, D.C., January 11, 1986.

7. President Kenan Evren was elected the seventh President of the Republic in November, 1982. Of this total only one individual has been a civilian (Celal Bayar). President Evren's term in office ends in 1989, at which time the Grand National Assembly must select a new President.

8. Emphasis added by author.

9. The Turkish Government provides no public data on linguistic, tribal or religious minorities, which legally do not exist. Figures given here must therefore be viewed as provisional, reasonably well-informed estimates.

10. The primary provinces in Southeastern Anatolia that comprise the Kurdish-speaking Turks are Bitlis, Diyarbakir, Hakkari, Mardin, Mus, Siirt, and Van. Today, this entire region continues to be under martial law. Clashes occur almost on a daily basis between armed Kurdish groups and Turkish security forces. Much of the military assistance that the Kurds receive comes from groups within Iraq, Iran and Syria.

11. All data within this section of this paper were provided by the Turkish General Staff or a subordinate command by order Gnkur. Bsk. Irginin 22175 C Mar 85 gun ve Gn. P.P.: 0050-186-85/Strj. Kv. Pl. (Strj. 6996) sayile mesaji. signed General Necdet Oztorun, Deputy Chief of the General Staff, 16 April 1985.

12. It is estimated that some 6 to 8 million speak Kurdish

and reside primarily in the above provinces. Today, this entire region continues to be under limited martial law.

13.   The eastern provinces where the Revolution took root finds only 10 percent of the Army cadets are recruited, whereas the percentage of the total population is 14.6 percent.

14.   On several occasions when consensus between political parties has been absent (e.g., 1960s), the existence of a well-organized bureaucracy has helped the political system survive conflict and indecisions.

References

Clapham, Christoph, George Philip.   1985.   The Political Dilemmas of Military Regimes.   London: Croom Helm.

Foreign Broadcast Information Service (FBIS), Vol. 7, September 17, 1980. Washington, D.C.: U.S. Government Printing Office.

Janowitz, Morris. 1960.   The Professional Soldier.   New York: The Free Press.

Janowitz, Morris. 1977.   Military Institutions and Coercion in Developing Nations.   Chicago: University of Chicago Press.

Legg, Keith.   1969.   Politics in Modern Greece.   Stanford: Stanford University Press.

McWilliams, Wilson C. 1967.   Garrisons and Governments: Politics and the Military in New States. San Francisco, CA: Chandler.

Nyrop, R.F. 1980.   Turkey: A Country Study.   Washington, D.C.: U.S. Government Printing Office.

Veremis, Thanos.   1978.   "Some Observations on the Greek Military in the Inter-War Period, 1918-35."   Armed Forces and Society 4: 527.

5

CHINA: A
DEVIANT CASE

Gordon Bennett

The PLA is an armed force which carries out revolutionary political tasks under the Party leadership.... (Su, 1985)

...In general, the security of a nation depends on the unity of its people.... (Beijing Review 1986:4)

To what extent do armed forces contribute to "social integration?" In China, no less than elsewhere around the developing world, if "social integration" is assumed to be reasonably progressive and important, then analyzing how well armies promote integration should help interpret how well they promote or retard social progress. Three possibilities occur:

Hypothesis 1, Military Leads: Developing country armies contribute to integration because they are well positioned. Where government is marked by personal loyalties, weak institutions, and instability, armies are the ultimate source of order and authority, and are responsible for good government. They overlap civilian administration because they accept internal security and constabulary duties. They develop political competence. Sometimes they act as a conduit for technology beginning with foreign military aid and the construction of native arsenals. Officers acquire managerial, engineering, and construction skills. Some governments go as far as to organize military production separately from the civilian economy. Military careers, and non-career military participation come to be perceived as a "modern" sector of employment with opportunities for upward mobility. Ambitious young officer recruits are less likely than are civilians to hold to religious, sectarian or communitarian values in their political thinking. Military elites are more likely to be Westernized, and more likely to have gone abroad for training. They are more likely to be recruited from middle- and lower-middle-classes than were their traditional predecessors who tended to come from upper classes or aristocracy. And they are more likely to have been recruited civil-service style by unbiased criteria (Janowitz, 1964).

Hypothesis 2, Loyalty Test: The premodern practice of recruiting armies from "backward" but "martial" groups who were numerically a minority, weak economically, low in sophistication, and dependent upon central authority for their status, has been replaced with modern reluctance to arm minorities whose loyalty is questionable:

> Modernizing a country's military has little to do with de-ethnicizing it...In fact, many civilian elites who are responsible for up-grading armed forces may shy away from just anyone in the society with such valuable skills. Civilian officials will only feel comfortable if they know that the men acquiring such techniques and weapons are wedded to the existing state regime by bonds of communal loyalty above and beyond the legalistic obligations of national citizenship. (Enloe, 1980:144).

Should Loyalty Test prove compelling, then developing country armies offer powerful incentives to minority communities to either demonstrate loyalty or move toward separatism.

Hypothesis 3, Military Retards: Worse even than remaining neutral, developing country armies slow down social integration. As social and economic development proceed and mature, the military's political importance falls. Military spending increasingly takes a back seat to development-related investment in education, technology acquisition, and export promotion. Military careers impress young people as carrying low status, and military participation becomes a lower-class preserve. Army units commanded to quell domestic disorder consequently yield a spectacle of the ill-educated clubbing and firing upon the severely alienated.

The Concept of "Social Integration"

Sociologists sometimes think in terms of a "social fabric" made whole and strong by thickly interwoven customs, values, institutions, and interests. Just as real fabric may not be strong uniformly, so may social fabric have weak threads or "tear lines" along which the first rips will advance. The concept is informal - I know of no refined, or statistical, integration scale as such - but it does point us toward asking where to look for the first tears. Inversely, social integration is strength across potential tear lines. For example, an American community where racial integration had progressed and reinforced racial fault lines would be correspondingly less likely to experience inter-racial conflict when pressured by an incident such as a white police officer shooting a black teenager, or a school board deciding to end crosstown bussing and revive neighborhood schools.

Ethnic tear lines receive the most attention probably because they are visible. Ethnic communities display remarkable tenacity in modern society: Spain has Basques, Canada Québecois, the United States a list of hyphenated Americans, Latin America Indians, Japan Ainu, and so on. And ethnic communities more sharply than other groups are set off by distinctive racial or physical characteristics, native languages, separate religious traditions, distinctive economic roles, or residential concentration.

Non-ethnic tear lines, though subtler to detect, can make a social fabric just as vulnerable. In modern China, non-ethnic faults include: (1) region--north/south and provinces and sub-provincial districts, all packaging differences toned by climate, economy, foreign contact, migration experience, political history, dialects, social life, arts, and cuisine; (2) development differential, separating coastal and riparian cities having elaborate ties to external commerce and cosmopolitan culture, from more provincial interior towns and villages where extra-local linkages remain largely primitive; and (3) social class, the line with blue collar and farmer on one side and white collar and bureaucracy on the other.[1]

With these hypotheses and concepts in mind, I will review Chinese data pertaining first to "ethnic" integration, then to "non-ethnic" integrations, and examine how both types are affected by China's military institution.

Ethnic Integration in China

"Ethnicity" in Chinese is zu (race, or nationality) and connotes to Chinese speakers one of 55 officially recognized groups that together make up China's population. Han zu (93 percent) overwhelm the other 54 "national minority" zu, the largest of which, Zhuang in south China, contributes a mere 1.3 percent.

China's "national minority" peoples, as Table 1 shows, are concentrated along sensitive borders with North Korea, the Soviet Union, Pakistan, India, and Indochina. They are more agricultural and less literate than the Han majority, and have lower life expectancy at birth. Their segregation from the Han people is captured by two demographic indices. Poston and Shu's geographic differentiation index GEODIFF measures distribution of a minority across China's provinces: if 0, the whole group lives in just one province; if 1, the group is distributed equally among all 29 provinces. China's largest minorities are shown to be significantly concentrated in a few, mostly border provinces. Poston and Shu's minority diversification index MINDIV applies to provinces instead of peoples: if 0, then only one minority group lives in the province; if 1, then all minorities live there in equal proportion.[2] Provinces with the highest proportion of non - Han population are arrayed along or near China's southern and western frontiers.

These observations merely open the story. Writers regularly slight the subject of minorities. Typical is correspondent David Bonavia, who devotes three brief pages to considering the potential of the "fringe cultures" (his term) to become important to China in the event relations with neighboring states should turn hostile. (Bonavia, 1982:286-89). At the scholarly level, intellectual historian Lin Yü-sheng simply equates "Han" with "Chinese." (Lin Yu-Sheng, 1979). Such treatment of ethnicity by analysts outside China has origins inside. Despite all positive, symbolic attention central

Table 1

China's Largest Minority Populations, 1982

	Pct. Pop. of China	GEO-DIFF (see text)	Pct. illit.	Pct. Agr.	Life Expec. (Yrs.)	Geog. Concen. b	Pct. Min. in(6)	MINDIV of (6) (see text)
	(1)	(2)	(3)	(4)	(5)	(6)	(7)	(8)
Zhuang	1.3	0.15	34	91	58	GXI'	38.5	0.21
Hui	0.7	0.91	43	62	55	Diffuse		
Uygur	0.6	0.002	44	86	41	XIN'	60	0.385
Yi	0.55	0.53	63	95	48	YUN'	32	0.8
						SIC	4	0.72
Miao	0.5	0.675	60	95	46	GUI	26	0.75
						YUN'	32	0.8
						HUN	4	0.73
Manchu	0.4	0.71	18	61.5	46	LIA'	8	0.5
						JIL'	8	0.55
						HEI'	5	0.6
Tibetan	0.4	0.69	75	92	39	TI'	96	0.01
						SIC	4	0.72
						Qin	40	0.62
Mongol	0.3	0.46	30	75	49	IM'	15	0.29
Han	93	0.95+	20-30	80	67			

'GXI=Guangxi, GUI=Guizhou, HEI=Heilongjiang, HUN=Hunan, IM=Inner Mongolia, JIL=Jilin, LIA=Lioning, QIN=Qinghai, SIC=Sichuan, TIB=Tibet, XIN=Xinjiang, YUN=Yunnan. ᵇ Border province

Source: Poston and Shu 1985, based on data from China's 1982 national census.

authorities shower on colorful national minorities, the continuing propensity of Hans to think of non-Han minorities as uncultured inferiors is an irrepressible parameter in everyday social life.[3]

If one conceives of ethnic integration minimally as citizenship - enjoyment by minority peoples of the same status, rights, and protections as Han people enjoy - then China's zu might be well on their way to formal integration. Law and policy now treat them as equal or even privileged (for example, minority peoples have been widely exempted from the one - family one-child restriction), however strongly age-old discrimination and segregation patterns might persist informally.

If, however, one conceives of integration more expansively as substantive equality - bringing minority zu into line with the Han people with respect to educational attainment, job and professional opportunity, income, and the like, including resorting to affirmative action of some type if necessary to make visible progress - then the record shows wide variation among groups. Koreans, Hui, Manchus, and Mongolians are more integrated in this sense than others (Poston and Shu 1985: 32-33). If one conceives of integration still more expansively as cultural assimilation - the gradual disappearance of highly visible peculiarities (communal schools, linguistic differential, religious observances, ethnic dress, etc.), however alive might remain less visible traditions (beloved ethnic humor, literature, songs, holidays, family customs, etc.) - Chinese authorities in the 1980s, in sharp contrast with a few years ago, actually encourage ethnic separateness; for example, Chinese family law in both word and fact now respects unique minority marriage and family values. Finally, if one conceives of integration as actual ethnic assimilation - inter-breeding to a degree that even physical traits are blended - the China's zu are integrated hardly at all; Manchus and Mongols excepted, Han intermarriage with non-Han is practically unknown.

If one insists upon a higher conceptualization, equality or a form of assimilation, then the Loyalty Test hypothesis unquestionably is consistent with the experience of the People's Republic. People's Liberation Army (PLA) recruitment still reflects majority disrespect of China's small minorities. If one is satisfied with the lower conceptualization of citizenship, then the applicability of an Enloe-type argument is possible but ambiguous.

Chinese conditions historically never corresponded to the central image in Enloe's argument - a ruling majority faced with unintegrated minorities who, if armed, might threaten the state. In China almost the reverse was true: for three centuries a ruling minority depended increasingly on a subject majority! The last imperial dynasty (the Qing) was formed in the early 17th century by a northeastern tribal chieftain (Nurhaci, of the Manchu Aisin Gioro tribe) and his sons (Abahai and Dorgon), who not only united other northern tribes but also attracted Chinese collaborators with the appeal of a strong, moral new state to replace the crumbling, rebel-ridden Ming state. The army of 169,000 that conquered the Ming state was about 49 percent Manchu, 21 percent Mongol, and 29 percent Chinese. Hardly a generation had passed before the privileged Manchu "banners" posted in fortified residential quarters about the empire began to prove less able than the already more numerous Chinese constabulary (the so-called "Army of the Green Standard"). Complained the Kang Xi Emperor of his Manchu generals, "men with the founding emperor's blood in their veins," after an 8-year campaign (1673-1681) to defeat a rebellion from the south:

> year after year, I had to watch them blunder and fail, hesitate to advance, stay snug in their base camps; and had to rely--even as rebellious Chinese generals cut the Manchus back--on other Chinese generals to turn the tide. (Wakeman, 1975:90).

By the 19th century the Manchu banners had become so useless the court had to commission Chinese viceroys to raise armies to suppress major rebellions. The Qing fell in 1911, and all subsequent Chinese governments were established and controlled by Han Chinese.

Whatever the message this history might bear for Loyalty Test, an apparently persuasive case for Military Leads can be built from contemporary data. I will proceed now to test it against Chinese evidence.

The PLA Role

The PLA since 1949 has been well-positioned to contribute to ethnic social integration. In the 1950s, early in the new Communist regime, it was contrived to be a vehicle for upward mobility, especially for poor and isolated rural farmers whose life chances were nil otherwise. It excelled at this function for many years. Many more youth wished to serve than were accepted. A young commune member selected for a three-to-five year stint could look forward to better education, likely Party membership if he (or in one or two percent of the cases, she) were invited to re-enlist, improved social status, and favorable treatment for family members while they, healthy offspring, were diverted from farm labor (Nelson, 1981:18-21). Over the years about 10 percent of eligible Chinese young people were recruited in the PLA.

Military participation helped one's subsequent career. By the early 1980s, some 500,000 soldiers per year were returning to civilian society, making a total of about 15 million demobilized since 1949. Most of this sizable contingent were given good jobs. Very often discharged veterans were assigned to spots away from their homes where their experience was felt to be needed. The considerable flow of discouraged Chinese refugees to Hong Kong over the years has been singularly lacking in veterans, suggesting a relatively high satisfaction rate among this group (Nelson, 1981:20-21). Thus some meaningful proportion of local elites in China owe their life chances to military participation.

Organizationally, the evolving system of military regions and multiple channels of authority helped institutionalize the army's responsiveness to civilian political needs, including subnational ones. Operational units are separated into "main force" units commanded and controlled from the General Staff Headquarters in Beijing, and "local" units under the control of local Party Committees. Both types are deployed throughout China under commands that roughly correspond to the country's administrative or local government boundaries (Military Region = several provinces, Military District = province, Military Subdistrict = special district, People's Armed Forces Departments, or "militia" = county or municipality). The Military Region (MR) administers all troops within its jurisdiction (that is the MR recruits, trains, recommends promotions, demobilizes, administers military justice, manages army-civilian relations, and supplies "its" main force units), although the General Staff Department in Beijing commands them, and the General Political Department in Beijing supervises the hierarchy of Communist Party committees and branches in the same units. The 37 or 38 main force corps of armies normally manned at about 46,000 in three divisions plus support units operate most of China's artillery, armor, and heavy equipment.

The MR, with its subordinate Military Districts, administers and commands the local independent regiments and battalions; intended for final defense of their localities, these units are smaller and more lightly equipped. Frequently local authorities assign them to assist with construction projects, planting and harvesting, supplementary security, disaster relief, and other local government priorities (Nelson, 1981:17). The line of command almost defies description insofar as it reflects an effort to balance the regional headquarters with the central high command. The navy and air force, as well as artillery, armor, and engineering combat support arms, all maintain liaison personnel in the MR's to try and reduce confusion. Should an MR's turf become an actual theater of combat, the center surely would grant it command authority.

However complicated this structure, the system afforded ample opportunity to promote social integration. Military commanders, or political and military authorities acting in concert, had only to choose to take advantage of it. Data showing overlap among military, Party, and government elites in the Communist regime's first two decades support such a "well-positioned" view; in recent years the three hierarchies have become disentangled (Whitson, 1973; Scalapino, 1972).

A high point of apparent military influence in social policy occurred in the early 1960s. The army was hailed as a political "model," and in 1963-64 a full-fledged "Learn from the PLA" campaign was launched to persuade civilian bureaucracies to follow military solutions to deteriorating political control. These "solutions" have a revealing history. During the Communists' first decade, Party membership predictably grew, from 4.5 million in 1949 to 14 million in 1959. Intra-Party conflict followed closely as Chairman Mao and his allies determined to "continue the revolution" disputed ever more bitterly with other senior leaders no less determined to postpone ideals that threatened visible gains in stable economic growth and political order. A simultaneous thrust to modernize China's revolutionary army along Soviet lines eroded political controls and led Mao to a critical 1959 showdown with his Minister of National Defense, Peng Dehuai. Peng was replaced with Marshall Lin Biao, who set in motion a multi-pronged initiative to repoliticize the PLA (Gittings, 1967). Mao was impressed enough not only to launch "Learn from

the PLA" but also to demand the creation in civilian
bureaucracies of PLA-like "political departments." Manned by
PLA personnel or demobilized soldiers, these new departments
competed with (and of course were much resented by) Party
committees already in place in these agencies (Bennett 1973).
On the eve of the Cultural Revolution, local army units also
were assigned a prominent role in the political socialization of
school children, apparently with no small effect. By one
account of a student at a vocational high school in 1965:

> We spent a lot time with the PLA. Between militia lessons,
> exchange visits, joint athletic meets, and joint recreation
> parties, we got to know many of the soldiers very well.
> An Air Force unit was stationed nearby, making exchanges
> quite convenient. We would mingle with the troops, ask
> them questions, and try to emulate them. We all had great
> respect for the PLA. (Bennett and Montaperto, 1971: 25).

With capital like this, the potential for the military to be a
vehicle for ethnic integration would appear to have been
substantial. Before concluding Military Leads should prevail,
however, some contrary evidence must be weighed.

Upward Mobility. As far as I can determine, while the
PLA served Han peasants well as a vehicle for career and
social mobility, it served minority communities less well if at
all. Good data are lacking, but anecdotal evidence suggests
that minorities were brought into the army primarily to serve
local control purposes. They were recruited into "local"
regiments and battalions far more frequently than into "main
force" divisions. In contrast to a common pattern for their
Han counterparts, they were regularly demobilized to their
home place instead of being assigned to overcome a shortage
of local cadre talent elsewhere.

Data are completely lacking on how minority PLA recruits
experienced their military participation, a question begged by
the context of the PLA's general "internal colonization" mission.
Knowing this would be extremely helpful.

Politically Responsive Army. No image of Chinese politics
is more durable than the Communist party-army, the armed
"fish" swimming in an "ocean" of popular support, suggesting a
military force sharing civilian political goals. The image may

have lost its lustre during 1954-59, while Defense Minister Peng pushed for Soviet-style military professionalization, but afterwards it shone again. Not until the infamous Tiananmen Massacre of June 4, 1989 did the myth of an army of the people finally explode.

A problem in viewing the PLA and Communist Party as institutional allies in China is the dominant cyclical Left/Right pattern in politics, which the PLA did not follow. The Party Left followed a political change agenda; its slogans were "class struggle," "dictatorship of the Proletariat," and "continuous revolution."   The Party Right followed a state-building, economic development agenda; its slogan were "new China," "joint dictatorship of all revolutionary classes," and "united front":

> Like the People's Democratic United Front within the Han nationality, the united front of the different nationalities, with the Han nationality as its core, has steadily expanded and developed in the course of the Chinese revolution. As in the past, this united front still consists of two alliances. The first alliance is that between <u>the Chinese working class and the laboring people of each national minority</u>. This alliance is the foundation of China's revolutionary united front of all nationalities; on the foundation there exists the alliance between <u>the laboring people of all of China's nationalities and nonlaboring people among the national minorities</u>. (Mosely, 1966: 130).

While much Chinese politics has revealed a tug of war between Left and Right agendas, the PLA--because of its revolutionary traditions, low technical level, Lin Biao's political direction, and Mao's leadership--lined up more often with the Left agenda. It was more an army of the Party Left than a simple Party army. Military influence rose and fell with Left influence. On the question of ethnicity specifically, the Left strayed not at all from Lenin's view that ultimately the only way to overcome "nationalities inequality" and "nationalities oppression" is for class consciousness and socialist society to rise in salience over ethnic consciousness. Thus the army grew identified with the intolerant view that national and ethnic loyalties are counter-revolutionary: by this view, while non-Han peoples might be accommodated occasionally, as a tactical means, such accommodation in the long run saps energy from revolutionary ends.

Legitimation Through Disaster Relief. China regularly rushes PLA rescue teams, fresh water trucks, and medics to natural disaster sites. International relief assistance was neither solicited nor accepted until 1988 (China Daily, 18 October 1986: 1). Army vehicles and encampments emblazoned with slogans make explicit the source of welfare activism - central political authority. In Chinese political culture, people are conditioned to view central government as a parent-equivalent benevolent in situations of need. On the one hand, PLA disaster relief affirms central government legitimacy, and so has potential to contribute to social integration. On the other hand, major relief episodes under the People's Republic have been limited to Han areas. Little relevance to Han-minority relations is discernable.

PLA Economic Construction. By official account, the army actively helps minorities develop their economies and cultures. After securing Xinjiang after 1949, large contingents of troops were sent "for reclamation," factory building ("using money saved from military spending"), and infrastructure construction. After Tibet was liberated in 1951, troops built interregional highways and opened an air line from Lhasa to Beijing; army efforts "removed the natural barriers between Tibet and the rest of China, and envigorated the economic, scientific and cultural life of Tibet" (Su, 1985:10). Nelsen argues in a similar vein that Han colonization of frontier regions coupled with farmland development and other infrastructure construction activities by PLA "Production and Construction Corps" helped integrate those regions into China economically (Nelson, 1981:19, 192-93).

This account is impressive as far as it goes. The PLA "Railway Corps" and several "Production and Construction Corps" contributed massively to the economies of border regions - farming, reclaiming land, constructing water control and irrigation works, building railroads, highways and airports, and opening factories. Today the Xinjiang Production and Construction Corps still accounts for one fourth of the region's economy.[4]

Still, a military contribution to ethnic integration has been lacking. However much the PLA might have contributed to economic, political, and defense integration - by making the minority-populated areas more prosperous and productive, by extending Beijing's authority to remote border regions, and by fortifying China's international frontiers - the occupation of minority-populated regions by largely Han forces, coupled with the use of those forces to discourage autonomous behavior by minority communities, has left minority leaders with no real alternative to submission.

PLA forces disbanded some two million enemy (Guomindang, or Republican) soldiers in campaigns throughout China's southern and southwestern border provinces - Fujian, Guangdong, Guangxi, Yunnan, Guizhou, Sichuan, and Xinjiang - between November 1949 and August 1950. Then, again by official account, "In accordance with the Agreement on the Ways for Peaceful Liberation of Tibet signed in May 1951 between the Central People's Government and the local government of Tibet, the PLA entered Lhasa and other areas of Tibet in December of the same year and helped emancipate the million serfs there" (Su, 1985: 10).

History failed to stop with "peaceful liberation," however. In the three largest minority-populated border regions, subsequent PLA hostility toward local leaders was greatest in Tibet, next so in Xinjiang, and least so in Inner Mongolia, which correlates inversely with level of ascending Han in-migration. Without Han settlement, the PLA alone did not promote integration.

Tibet has proven least hospitable to Han settlers, who amount to only seven percent of the region's population. Like a self-fulfilling prophecy, the new Chinese government after 1949 viewed Tibet as hostile - a penetrable border region twice the size of Texas, sparsely populated by people living under one of the world's most anachronistic social systems (no highways or railroads, economy of estates owned by monasteries and worked by serfs, government dominated by Buddhist lamas), anti-PRC maneuvers by "imperialist" foreigners - and its assertive military-backed policies to control Tibet prompted ever more hostility. As the years dragged on, Chinese pressure on the feudal Tibetan theocracy mounted, only to be answered by wider Tibetan armed resistance. By 1958 (a radical year in the Chinese political cycle), Chinese statements were denouncing religion and calling Buddha a reactionary, and Chinese troops were resorting to ever more coercive tactics against Tibetans, their monasteries, and their villages. Local resistance forces linked up in open rebellion, and the Dalai Lama fled to India the next year. By the early 1980s the number of monasteries had shrunk to 10, from 2,464 before 1959; journalist Fox Butterfield found a PLA company encamped on the razed site of a famous temple on Medicine King Hill in Lhasa (Butterfield, 1982:429-30). By 1989, multiple violent incidents had led to Beijing's reimposition of martial law.

Steady in-migration from China proper into Islamic Xinjiang raised the Han presence in the regional (now provincial) population from about five percent in 1949 to forty percent in 1982. Nonetheless, the pre-Communist history of anti-Han incidents projected forward. Perhaps fueled by the contemporaneous Tibetan rebellion, several anti-Han incidents broke out in 1958-59 that brought a PLA response: Wusu in September, Yining in November, and Hetian for two months (a crowd of thousands attacked a prison, freeing some 600 inmates). In the spring of 1961 about 4,000 Uygurs reportedly revolted against Chinese troops at Yining. Several tens of thousands of Uygurs and Kazak rioters were dispersed by gunfire in 1962; 62,000 crossed the border into the Soviet Union (McMillen, 1979: 119, 122-23, 241). In 1980, several hundred civilians and soldiers reportedly were killed during a fight in Aqsu, near Kashgar. Angry Uygur crowds roamed the streets shouting "death to all Chinese," and ultimately were brought under control by PLA divisions stationed in the region. Apparently this was only one of several such incidents involving gun battles and execution of suspected "rebel" leaders (Butterfield, 1982: 429).

The PLA occupied Inner Mongolia after the Japanese retreated in 1945, and organized it as the first minority "autonomous region" in 1947. Yet stimulated perhaps by Mongolia's location - proximate to the capital Beijing, and straddling the Sino-Soviet border - a combination of heavy Han in-migration and Han-Mongol intermarriage has produced the odd result that the population of the present-day (considerably smaller) Inner Mongolian Autonomous Region, or IMAR, is only about five percent ethnic Mongol. The IMAR does not have the same history of suppression of local native political aspirations as do Tibet and Xinjiang. Limited ethnic cleavage accompanies little evidence of a PLA role.

Minority communities in smaller jurisdictions are less studied than are Tibetans, Uygurs, and Mongols, yet problems there still surface from time to time. For example, the "Shadian Incident" in 1975 involving the small Hui settlement in Yunnan Province never was officially reported. The local Party Secretary, evidently frustrated over successful Hui resistance to her demands they abandon their separate religious practices, even after years of intimidation, ordered bleeding pigs' heads thrown into their courtyards and wells to contaminate them. When the Huis resisted, she called in a regular army infantry division to annihilate their entire town, leaving few survivors. (Heng and Shapiro, 1986:115-18).

Summary. In this section focusing on ethnic social integration, I find the Enloe Loyal Test hypothesis to fit China only ambiguously, since Chinese military history does not conform to the "tradition" part of her argument. And although I find the Janowitz Military Leads hypothesis doubtful in the case of China, the PLA is well positioned to promote ethnic integration. Unfortunately direct evidence proves unfriendly: the PLA offers upward mobility only to Han youth; the PLA has been less of a broadly popular army than an institutional member of the Left coalition at the center; PLA disaster relief primarily serves Han areas; and PLA contributions to economic construction in non-Han areas give more an appearance of "internal colonization" than of integration.

Non-Ethnic Integration

As small as the military's role in ethnic social integration may be, there remains a meaningful possibility of army contributions to non-ethnic integration across region, development differential, and social class. Even beyond that remains a possibility that post-1980 military reforms might greatly alter the PLA's contribution (I call this the New Vanguard hypothesis).

The PLA Role

Region. Historically Chinese society was remarkablely decentralized, and locally-oriented; for most people capital and emperor, tucked away in China proper's northeast corner, were distant physically and culturally. Indeed, a leading explanation for the survival for two millennia of one Chinese state territory is the flexibility of a traditional state unified by such strands as a single hierarchy of administrative authority, a single orthodox ideology, and tax deliveries to the center. All else fell in the realm of local affairs. This impressive, durable political culture did not disappear in 1949. Leading characterizations of China's economy and society under the Communists notably emphasize its "cellular structure" or "nested local hierarchies," both implying a significant degree of local autonomy.

I have yet to encounter data on the proportion of PLA recruits who serve in their home province. But at elite levels personal networks of patronage and obligation if not outright military "factions" formed around the regional commands, parallelling tendencies in civilian administration. No less than five early-1980s military reforms were targeted at personalistic, regional power bases: (1) up-or-out promotions, by age 30 in the case of company commanders; (2) rotation of military region commanders; (3) mid-career reassignment of field-grade officers, nearly all of whom now must attend advanced or specialized officer schools, to erode vertical cliques; (4) having unit members evaluate candidates for promotion, hoping that "democratization" would undermine familiar sponsor-protégé promotions; and (5) annual written exams as part of the promotion process (Nelson, 1981: 228-29). As this reform thrust testifies, the army is no less afflicted than the rest of Chinese society with regionalism.

Development Differential. China's economic modernization in the 20th century was led by, if substantially restricted to, coastal and riparian cities having direct diplomatic and trading ties to Japan and the West. Little "true modernization" took place in Republican China (1912-49); no inter-regional freight system - by river steamer, railway, or all-weather highway - developed: "more goods were carried in more carts to more shops in more markets more frequently convened - but there was no systemic change" (Skinner, 1964). The wide gap between more developed coast and more primitive interior the Communists inherited became the subject of explicit "locational policy," first favoring the interior, from 1956 building on coastal strength, then gradually favoring the interior again.

Drawing personnel from less modern sectors, the PLA has failed to bridge the gap. Reform-minded efforts to upgrade PLA personnel and especially officers through selective recruitment, promotion, and improved training all bear witness to "shockingly low" education and skill levels (Nelson, 1981:230). Ground forces dominate the PLA, infantry dominate ground forces, and peasants dominate the infantry. Only in 1983 did the Director of the PLA General Political Department (Yu Qiuli) publicly recommend that officer recruitment be shifted from infantry to more urban and better educated specialized service arms (Jencks, 1964: 60). For the most part, the army has remained a reservoir of rural backwardness.

Social Class. Humble classes of the *ancien régime* - "workers, peasants, and soldiers" - were made the privileged classes of Mao's new China. But if privileged status was now the PLA's, so was reaffirmation of a popular cultural attitude: soldiering was equivalent to the social position of workers and peasants. Hao tie bu zuo ding; hao ren bu zuo bing ("Good iron is not wasted on nails; good men are not wasted as soldiers").

Moreover, after the revolutionary war years, nationalistic motivations to join the military gave way to life-chance motivations. Once wartime enemies were gone, poor rural youth facing rather dismal futures volunteered only to improve their lot. The change from nationalism to careerism further opened the PLA to shifts in careerist thinking. As the life-chance value of other life pursuits rose over military participation, which happened increasingly after 1978, volunteers thinned. No part of this story is friendly to believing the army was active as a reliable agent of social integration across classes.

"New Vanguard"

This final hypothesis asserts that all previous interpretations must be abandoned; because military reform since 1980 is a model for all reform, the PLA could help integrate China socially to a degree the Maoist military could not.

> The PLA has lost much of its political power on questions not related directly to national defense. The new social role of the PLA is an institutional example of the way the reformers would like to see other Chinese institutions reformed, and the PLA's economic role now provides very concrete support to the modernization of the Chinese economy (Bullard and O'Dowd, 1986: 711).

Should this thesis prove correct, it would favor Military Leads, certainly for non-ethnic integration and perhaps for ethnic integration as well; a vanguard modernizing PLA offering new professional educational opportunities might attract recruits whose careers would be less bound to old cliques. But is it correct? I think the jury is still out.

Bullard and O'Dowd draw heavily from announced new policies, Deng Xiaoping speeches, and military elites' declining political power in the 1980s. These are important developments but not for determining the military's role in social integration. Continuing progress of China's post-1978 reform program will be a function of several factors: experience with the new rules of economic behavior, inherent strengths and weaknesses of the emerging half-socialist half-capitalist bureaucratic system, the outcome of several new policy experiments, favorable movements on relevant international markets, strength of domestic political supporters and opponents of the reform direction, top leadership's continuing commitment to Deng's reform goals (he first hoped to retire in 1987), and the level of economic performance sustainable under the reform system. This whole complex movement might profoundly affect the future civilian role of the military, a future not yet visible.

Summary. In this section, focusing on non-ethnic integration, I find nothing to support <u>Military Leads</u>. 1980s reforms were directed at controlling regionalism in the PLA. The 1980s PLA remained a reservoir of rural backwardness. And professionalization to a degree that might be expected to attract intellectuals to volunteer has yet to happen. 1980s military reforms themselves have yet to progress far enough to change the rules.

Conclusion

Chinese data found strong reason to doubt both <u>Military Leads</u> (represented by Janowitz) and <u>Loyalty Test</u> (represented by Enloe). I must conclude <u>Military Retards</u> is most persuasive by default, at least for China. Since the work of Janowitz and Enloe is widely accepted, I judge China to be a deviant case. Why that should be is unclear. I would cast first in the shoals of political culture. Several Chinese institutions have exhibited either constancy or consistent directions of change over long numbers of centuries, whether the ruling dynasty was native (Han, Tang, Song, Ming) or conquest (Yuan, Qing). Military policy was derivative, not formative.

Notes

1. Often this cleavage is expressed in Chinese rhetoric as "worker" vs "intellectual" and accordingly has generated a stream of "policies toward intellectuals."

2. See Posten and Shu, 1985. Both GEODIFF and MINDIV are defined as where X is the population size of one minority group in one province. In 1988 Hainan Island, formerly part of Guangdong Province, became China's 30th province.

3. Uygur <u>zu</u> members of a Chinese "arid lands" delegation to the U.S. in 1985 were routinely shunned by their Han colleagues. The Uygurs roomed together, and despite holding

professional credentials they rarely joined in discussions. My own visit to the Uygur home region in 1985 left similar personal impressions. Our mostly Han hosts described the local Uygurs as easy going, fond of singing and dancing, and famous for growing tasty melons, a characterization hauntingly reminiscent of the indelible American white stereotype of the post-bellum southern black.

4. The Xinjiang Corps. is credited with reclaiming over 10 million mu (1,666,667 acres) of farmland, 20 percent of the total now under cultivation. Its 169 farms and 729 industrial enterprises accounted for one fourth (2.6 billion yuan) of the total regional product in 1984. Most recruitment to the Corps occurs outside the region. In 1985 the Corps had "more than a million staff workers," and "a population of 2 million" on its farms and ranches. The Xinjiang Corps now accounts for over 15 percent of the region's population and 25 percent of its economy. It commands an entire gallery in the Regional Exhibition Hall. (Based on materials supplied by regional authorities during the author's visit in 1985).

References

Bennett, Gordon. 1973. Activists and Professionals: China's Revolution in Bureaucracy 1959-1965. A Case of the Finance-Trade System. Madison, WI: Department of Political Science, unpublished doctoral dissertation.

___ and Ronald Montaperto. 1971. Red Guard: The Political Biography of Dai Siao-ai. New York: Doubleday.

Bonavia, David. 1982. The Chinese: A Portrait. New York: Pelican Books.

Bullard, Monte, and Edward C. O'Dowd. 1986. "Defining the Role of the PLA in the Post-Mao Era." Asian Survey 26 (June).

Butterfield, Fox. 1982. Alive in the Bitter Sea. New York: Times Books.

Enloe, Cynthia. 1980. Police, Military, and Ethnicity: Foundations of State Power. New Brunswick, NJ: Transaction Books.

Gittings, John. 1967. The Role of the Chinese Army. New York: Oxford University Press.

Heng Liang and Judith Shapiro. 1986. After the Nightmare. New York: Knopf.

Janowitz, Morris. 1964. The Military in the Political Development of New Nations. Chicago: Phoenix Books.

Jencks, H. W. 1964. "Ground Forces", pp. 54-76 in Gerald Segal and William Tow, eds., Chinese Defense Policy. Urbana, IL: University of Illinois Press.

Lin Yu-Sheng. 1979. The Crisis of Chinese Consciousness: Radical Anti-Traditionalism in the May Fourth Era. Madison, WI: University of Wisconsin Press.

McMillen, Donald. 1979. Chinese Communist Power and Policy in Xinjiang, 1949-1977. Boulder, CO: Westview Press.

Mosely, George. 1966. The Party and the National Question in China. Cambridge, MA: Massachusetts Institute of Technology Press.

Nelson, Harvey. 1981. The Chinese Military System: An Organizational Study of the Chinese People's Liberation Army, second edition. Boulder, CO, Westview Press.

Poston, Dudley, and Jing Shu. 1985. "The Distribution and Composition of the Major Minority Groups in China." Austin, TX: University of Texas, Population Research Center Paper #7.005.

Scalapino, Robert. 1972. "The Transition in Chinese Party Leadership", pp. 67-148 in Robert Scalapino, ed., Elites in the People's Republic of China. Seattle, WA: University of Washington Press.

Skinner, G. William. 1964. "Marketing and Social Structure in Rural China, Part 1." Journal of Asian Studies 24, 1 (November): 22-55.

Su Wenning. 1985. China's Army: Ready for Modernization. Beijing: Review Press.

Wakeman, Frederick. 1975. The Fall of Imperial China. New York: Praeger Press.

Whitson, William, with Chen-hsia Huang. 1973. The Chinese High Command: A History of Communist Military Politics. New York: Praeger Press.

6

THE MILITARY AND NATIONAL INTEGRATION IN INDIA

Raju G.C. Thomas and Bharat Karnad

In many Third World countries, the military has played an important and forceful role in their domestic politics, economics and society. Indeed, the essential functions of the military in many Latin American and African countries has not been in the international arena deterring or resisting armed attacks on the state, but in the domestic arena dealing with armed insurgencies and acts of terrorism resorted to by dissident ethnic or tribal groups. The internal security role of the military has often led to military coups and the perpetuation of praetorian states (see Finer, 1976; Nordlinger, 1977; Perlmutter, 1977; Simon, 1978; Lowenthal and Fitch, 1986; Cohen, 1984). In South Asia, the military has been the dominant actor in Pakistan and Bangladesh. On the other hand, despite its size and power, the military has played a much lesser role in the Indian domestic political arena. In general, the principle of civilian control of the military has prevailed, and the military's incursions into the domestic politics of India have been only at the invitation, and under the control, of the civilian authorities.

Although the military has remained in the background in India, its large size and effective organization, its demand for control over considerable resources, and its potential ability to intrude into the domestic political arena, make it a significant force in national life. In a diverse and multi-ethnic society

127

such as India, the recruitment policy of the military and the location of its defense production programs are cause for competition and concern in the different regions. The military may be perceived also either as a threat to various ethnic groups seeking greater autonomy, or as the protector and unifier of the nation.

With some qualifications, the military in India may be viewed as an integrating force in its role as the defender of the nation against external threats, and as a showcase for secular, non-caste and non-religious values that Indian leaders have sought to impose on it since independence. But it may be viewed as a disintegrating force when dealing with problems of internal security, especially to the extent that it has retained its colonial constabulary structure and mindset.

Compared to most Third World states, the Indian experience has been unique because of continued civilian political control of the military within a Western parliamentary-type democratic system. Unlike Pakistan and Bangladesh, the military has remained apolitical. Defense policies to deal with external threats are largely made by civilians. The extent to which the military is deployed in counter-insurgency and anti-terrorist campaigns within the state are again determined by civilians. On the other hand, both Pakistan and Bangladesh have experienced the overthrow of their civilian regimes and the seizure of political power and authority. The military regimes in Pakistan and Bangladesh have claimed that their intervention into the political sphere prevented the disintegration of the state.

One of the reasons for this happy situation in India is because the military, and especially the army, has always been given a broad-based and socially prestigious identity in the country. At the same time, the high-minded acceptance of civilian overlordship combined with generally good-natured resistance by the officer corps to civilian attempts at social engineering within the military, has done much to dissipate the tensions that might have arisen otherwise from the relationship. This has resulted in nearly 40 years of amicable civil-military relations. On balance, therefore, the military in India may be seen as an integrating force and as a defender of democracy.

On the other hand, the size, growth and potential power

of the military and its frequent use for internal security purposes could undermine those prevailing democratic norms. When the military is used to deal with insurgency, terrorism and a general breakdown of law and order in the states, this is usually done through various ordinances and amendments to the constitution that are passed by the ruling government that in effect bypasses democratic and normal judicial procedures. Moreover, the use of the military for internal security through such authoritarian measures has usually aggravated the fissiparous tendencies among the various ethnic groups in India. Rather than being seen as the symbol of unity, it may be seen as the obstacle to rectifying political injustices or achieving political freedom. Under these circumstances, the military in India may be viewed as a disintegrating force and a potential threat to democracy.

The Military as an Integrating Force

Until the Indian Army's assault on the Sikh's Golden Temple in the Punjab in 1984, the military had been viewed by most of the Indian population as a symbol of national unity. The integrating force provided by the military has been through its national recruitment efforts, its role as the defender of the nation against external aggression, and its role in providing disaster relief and other assistance to the civilian authorities during natural calamities.

Recruitment Policy. Although recruitment to the Indian army has tended to follow the pattern of recruitment by the British in the pre-independence era based on the concept of the "martial races" of northern India, the continuation of this policy after independence in 1947 has been partly due to historical accident rather than deliberate design, and partly due to an unstated "evolutionary" policy followed by the senior commanders aimed at minimizing dislocations and breaks with British tradition (Cohen, 1971; Kavic, 1967).

The earlier British rationale for such communally-separated regiments was that the various ethnic groups in India had different customs, traditions and eating habits that could not be easily integrated. It was also argued then that communal regiments produced a better fighting unit and a corps d'esprits. But these arguments may be misleading since no such efforts were made to separate the Indian ethnic groups in the Royal Indian Air Force and Royal Indian Navy. No doubt, these two services were much smaller than the British Indian Army. Nevertheless, the reasons for communal regiments may have had much to do with the policy of "divide and rule" rather than concern for the special interests of various ethnic groups (Thomas, 1986). Clearly, an integrated British Indian Army could have posed a threat to the empire itself. On the other hand, when separated and commanded almost exclusively by British officers before the Second World War, the British Indian Army not only did not constitute a threat to the empire but was effectively wielded to defend it against rebellious forces within India and in Britain's overseas campaigns in Asia, the Middle East, Europe and North Africa.

While the army was conducive to such communal division and representation in the formation of regiments, the air force and navy were not conducive to such communal separation. It would have appeared absurd to create a naval destroyer full of Sikhs or a submarine full of Gurkhas or a combat air squadron of Marathas only. The policy of divide and rule would have been too obvious if applied to the air force and navy. To be sure, the RIAF before the Second World War was only in a nascent stage, air power itself being of recent development. Given the small size of the RIAF, the British could perhaps argue that communal separation was unnecessary. Likewise, Britain's Royal Navy commanded the Indian Ocean and the need for a large Royal Indian Navy organized into communal units may be argued again as being unnecessary. Yet the threat from a potentially large and communally integrated British Indian Army was ominous and this may have been an important reason for communal separatism in the regiments. The historical importance of the Great Indian Mutiny of 1857-58 was not lost on the British when Hindu and Muslim soldiers of the British East India Company revolted against their British masters. It is interesting to note that the small but integrated Royal Indian Navy mutinied against the British in 1946, accelerating the move towards independence a year later.

The British legacy was difficult to reverse in the short run. Quite apart from the understandable resistance to change by priviliged ethnic groups within the army, there is the military's overarching concern with maintaining combat effectiveness and, therefore, an unwillingness to sacrifice the qualities of cohesion and esprit de corps of the fighting men especially in the infantry and armored divisions. The need to generate such qualities based on caste and community in the front-line regiments is strongly believed by the military high command (personal interviews).

So the main Indian infantry regiments remain largely Sikh, Dogra, Jat, Maratha, Mahar, Kumaoni, Garwahli, Rajput or Gurkha. And many old armored units, Skinner's Horse, 17th Horse, Deccan Horse, and Central India Horse, to mention a few - still have Sikh, Jat, Dogra, Maratha, and Rajput tank squadrons. In fact, so determined has Army headquarters been to conserve tradition and keep the "sword" sharp, mechanized outfits raised after independence such as the 63rd, 64th, 65th, 70th and 71st Cavalry, all feature both "mixed" and separate Jat, Sikh, Dogra and Rajput tank squadrons along the lines of the older and more renowned units. Senior army officers say that this determination is reflected, for example, in the continued support for the institution of Honorary Colonel of the Regiment, an honor that was part of the British era in India (Sinha, 1980: 121). Usually, the Honorary Colonel, usually a senior general, is expected to play "godfather" to the unit, lobby for better equipment, more personnel, and meet other regimental needs, and is expected to inculcate in the new regiment "traditions" of the old regiment he served in as a young subaltern.

Once these traditions take root, they join and consolidate the larger corpus of "traditions" which the military jealously guards against civilian interference. But as a show of its intent to move gradually towards the "integrationist" ideal, the Indian Army has acquiesced to the induction of "outsiders" into many previously "pure" regiments. Thus, Gujeratis - labelled a non-martial race by the British - have been permitted to join the Rajputs. Several Mahar regiments now comprise non-Mahars. However, the most conspicuous move in this direction has been to raise the only new infantry regiment in the last four decades, viz., the elite Brigade of Guards (BOG). Emphasizing this showpiece nature of this regiment, it was cobbled together by transferring the "best" battalion from each of the established regiments and forcing these disparate fighting groups to merge their identities.

The Brigade of Guards is undoubtedly a public relations success. Whether or not it is an operational success is still not clear, but it is significant to note that the higher echelons of the Indian Army are tentative and ambiguous about its effectiveness, choosing to refer to this by now 38 year-old regiment as an "experiment." In any case, the "cautious" policy of retaining the exclusive structure and character of the infantry and armored units, almost in accordance with the 1902 blueprint set out by Lord Kitchener, the then Commander-in-Chief of the British Indian Army, is a compromise arrived at by military men keen to maxmize "elan" and operational effectiveness by pulling one way, and the politicians looking to make the army fully representative and to enlarge the employment opportunities for their constituencies in the military by pulling in the opposite direction.

The compromise permits the fighting arms to retain their "combat edge" through selective recruitment policies, but with an informal "understanding" that should the strength of the infantry and armor be increased, the addition of new battalions to the old regiments will be on the basis of absorption of recruits from a larger pool of aspirants. Hence the presence of Gujeratis in the Rajput regiment and non-Mahars in the Mahar regiment. Politicians over the years have been content with this arrangement because of their respect for the military's sensitiveness and their unwillingness blatantly to trespass into areas of the Army's prime professional military concern about having superior fighting formations, by dictating who should or should not be recruited and in what numbers. The reluctance of the civilians to interfere is also because the Army is generally quite mixed, and getting more integrated by the year. An officer of the rank of a general explained the trend this way:

> "Infantry and armor have a proportion in the overall Army set-up. Then there are the supporting and service arms like artillery, signals, supply and ordnance, transport,

education, which are mixed units. What happens is that the proportion of those other than infantry and armor is larger. So when the expansion takes place on this broader base, the accretion of the traditionally non-martial classes is always greater" (personal interviews).

Communal Representation. Reversing British policy of priviledged communal recruitment and communal formation of regiments has not been easy for the government of India, for it implies reducing the opportunities for employment of those communities that had traditionally contributed large numbers of soldiers to the Indian army. The post-independence policy had become one of the sore points of the Sikhs, for instance, who under the British had contributed almost 20 percent to the pre-partition British Indian Army (another 30 percent was Muslim, mainly Punjabi, all of whom were transferred to Pakistan after its creation. The rest were Hindus, including Gurkhas). Decades later, even after partition and the dispatch of the Muslim units to Pakistan (constituting about 30 percent of the British Indian Army), the Sikhs formed only about 8 percent of the Indian Army. This contribution is likely to decline even further as the Indian government continues efforts to create a military that more fairly represents all regions and communities in proportion to their numbers in the Indian population. The Sikhs constitute only 2 percent of the total population.

Although efforts to create proportional representation in the military may cause resentment among those groups that were once favored by the British, the new Indian policy should gain favor with the majority of the rest of the population who earlier fell outside the historical and traditional sources of military recruitment. Thus, Sikh alienation in the recruitment process may be overcompensated by favorable Hindu reactions elsewhere in the country, especially given the present phase of Hindu revivalism and fundamentalism. After all, Hindus constitute the overwhelming majority of the population (80 percent) and in a democratic system, pampering the majority could not hurt any ruling party seeking votes to maintain their power, whether Congress, Janata or some other party aspiring to the win the next general elections.

Once a sensitive but now a receding factor has been the recruitment of Muslims into the Indian armed services. After partition, virtually all the Muslims, both officers and jawans, numbering about one-third of the total British Indian Army, were transferred to the new state of Pakistan. The transfer of Muslims to Pakistan also applied to the Royal Indian Air Force and Navy, although a few Muslim airmen and seamen remained with the newly constituted Indian units, since these services were more communally integrated than the army.

After the 1947-48 Indo-Pakistani war, there was little development and growth of Indian military manpower although there was considerable purchases of military equipment for the army and airforce to counter the weapons acquisitions in Pakistan through its participation in the American-sponsored Southeast Asian Treaty Organization (SEATO) and Central Treaty Organization (CENTO). However, even the annual recruitment to replace retiring military personnel did not attract many Muslims either because the recruitment process eliminated Muslims as security risks, or, as is more likely, because Muslims were deterred from joining the Indian armed services that were most likely to be deployed against Pakistan with whom they carried sentimental loyalties.

All the same, the Indian army, with the concurrence and approval of the government, has endeavored to act very professionally on this issue. When the 1965 Indo-Pakistan war began, a Muslim-majority battalion of the Rajput Regiment stationed in the crucial Poonch sector of Jammu and Kashmir state, far from being hastily withdrawn, was allowed to play its part in the execution of the Army's forward actions. According to several high-ranking Indian army officers, the fact that the battalion did not flinch and carried out its assigned role with considerable credit, sufficiently dispelled worry--at least within the military--about the loyalty of Indian Muslim soldiers (personal interviews).

On personnel matters too, the Army has attempted to act objectively, appointing and promoting officers on the basis of seniority and merit. Thus, the longest serving commandant of the National Defence Academy at Khadakvasla in Maharashtra state--India's equivalent of a joint West Point, Air Force Academy and Annapolis--was Major General Enayat E. Habibullah, who held charge for over six years starting in the late 1950s. It was General Habibullah who began "Indianizing" the ethos of the officer corps. As commandant of the military's premier officer training institution in its early years, Habibullah made it mandatory for gentlemen-cadets to acquire knowledge of Indian culture by requiring they attend Indian music and dance concerts and poetry recitations. He introduced the serving of Indian food at the mess, and otherwise forced the prospective officer cadre to appreciate the milieu for which they had taken the oath to defend. Later, Major General Shami Khan, a member of the Indian nobility of Rampur (and hence a relation of retired Pakistani General Sahabzada Yaqub Khan who was military governor of East Pakistan and later foreign minister of Pakistan), was given several postings on the West Pakistani front during his career, and eventually attained the rank of Deputy Adjutant General.

In the aftermath of the breakup of Pakistan, the creation of Bangladesh in 1971, and the apparent discrediting of Mohammed Ali Jinnah's "two nation" theory (namely, that Hindus and Muslims on the subcontinent constituted two separate nations), the political climate for entry of Indian Muslims into the armed forces has further eased. Adherence to the Islamic faith could not prevent the disintegration of Pakistan. Although only marginal, there are now more Muslims in India than in Pakistan, with Muslims in Bangladesh only marginally exceeding that in India. The prospect of another war with Pakistan has also receded since the late 1970s, partly because of India's growing military dominance of the subcontinent. Under these newer circumstances, the Indian military may have begun to attract more Muslims (figures for Muslim enrollment or the strengths of various castes and communities in the numerous regiments is considered information pertaining to the order of battle and, therefore, is classified).

External Defense. The wars against Pakistan and China have generally tended to produce a sense of national unity and purpose. This observation is probably more true during the war with China in 1962 than the wars with Pakistan in 1947-48, in 1965 and 1971. Sino-Pakistani relations generate more political sensitivity and emotion, confuse and strain the loyalty of Indian Muslims, and correspondingly increase the mistrust carried by Hindus of Muslims in India. However, apart from the 12 percent Muslim minority in India which in 1986 probably numbered about 94 million, Indo-Pakistani wars have produced a high sense of nationalism among Hindus, Sikhs and Christians, and perhaps also many Muslims who never supported the concept of Pakistan and who feel that their future lay with India.

Especially during the 1965 and 1971 Indo-Pakistani wars, the government of India went to considerable pains to highlight the fact that the war with Pakistan was between two different political systems - a democratic and secular state versus an authoritarian and theocratic state. The courage, bravery and sacrifice made by the few Muslims that served in the army, air force and navy for Indian secularism and democracy have usually been given a great deal of publicity. For instance, the bravery and death in the line of duty of a jawan in the tank regiment named Abdul Hamid was given much fanfare by the Indian press, much of this having been orchestrated by the Indian government which constituted the main source of information for such matters.

The 1962 war with China produced a greater feeling of nationalism and unity. Unlike the wars with Pakistan which continue to appear more like a domestic Hindu-Muslim communal conflict "with armor," as one American author put it, the war with China could readily unite all communities in India (Cohen, personal communication). Unlike the people of the subcontinent who have a common racial, cultural and even religious heritage, the Chinese are a different group of people. To be sure, there were tensions and misgivings among some Indian communists who had always looked to the external communist movement for inspiration and guidance. Mao Zedung and the Chinese communist revolution carried greater relevance for the Indian communist movement since economic conditions were similar in India and China.

However, at the time of the October-November 1962 Sino-Indian war, the Sino-Soviet split was already intensifying and communist movements all over the world were being split into pro-Chinese and pro-Soviet factions. This was to occur in India as well towards the end of 1963 following the open rift between the Soviet Union and China in July 1963. During the Sino-Indian war, several suspected pro-Chinese Indian communists, especially in the states of Kerala, West Bengal and Andhra Pradesh, were arrested under the newly reconstituted Defense of India Rules.   But the numbers were few and they were released following the termination of hostilities.   In short, political confrontations and wars against China and Pakistan (to a lesser extent) would appear to be good in the short run for national integration.   Indeed, some opposition leaders have alleged that such tactics were being employed by Prime Minister Indira Gandhi in order to gain political unity and support and to perpetuate Congress government rule in India.

Similarly, her son and successor, Rajiv Gandhi, has alluded to the "foreign (read Pakistani) hand" in the Punjab troubles and to "war preparations" across the border with an eye possibly to drumming up support for his party among the electorate.   These statements, allegedly in response to Pakistani reactions to massive Indian war exercises in early 1987 in the Rajasthan Desert, were made (not coincidentally) on the eve of state elections in three states. They were not politically productive.   Rajiv Gandhi's Congress party lost to communist parties in two states that year, West Bengal and Tripura, and "piggy-backed" to victory as a junior coalition partner in a third state, Kashmir.   It indicates the decreasing returns to Indian politicians from stoking paranoia, especially about Pakistan, which is increasingly perceived as less and less a conventional military threat.   Whatever the politician's motives for sounding periodic alarums, the military virtually ignores them, interpreting them, as do most politically sophisticated segments of society, as essentially political ploys.

Non-Security Aid to the Civil Authorities. As elsewhere in the world, one of the important functions of the military in India is to provide aid to the civilian authorities, especially during times of natural disasters. The Indian Army has carried out this task almost routinely by providing rescue operations and assistance to victims of floods during heavy monsoon rains. The Indian Air Force has provided air transport during such rescue missions. The domestic non-security role of the armed services is good public relations and projects the military as a friend of all Indians of every class, creed and community. Since the personnel from the armed services that are utilized in this role are drawn from all over the country, a sense of national unity and support for those less fortunate in the "hour of need" is felt among all groups of Indians.

While the military's role in providing disaster relief is generally taken for granted, there have been some suggestions that during prolonged periods of peace the armed services should be utilized on a more regular basis in India's civilian development programs. An extension of this argument is that a much larger "people's liberation army" should be formed along the Communist Chinese model whose functions would be both defense against external aggression during times of war, as well as participation in various economic development projects during times of peace. In this way, a large standing military would not prove to be an economic liability to the nation, and the greater interaction of the military with the people would keep the military in touch with the mainstream of civilian life.

Although the military has no objections to its role in providing disaster relief, it has strongly resisted any attempt to engage it in regular development projects (Thomas, 1986). From the standpoint of the military, such a policy would politicize the armed forces and reduce the professionalism and combat effectiveness of the armed forces especially if hostilities were to break out suddenly. Military training and alertness needs to be maintained at all times, otherwise slack and ineptness would creep in with disastrous effects for defense against external attacks. Even the decision by the first Indian Commander-in-Chief, General K. M. Cariappa, to commit the army to Prime Minister Jawaharlal Nehru's proposal, namely, to use the vast Army-held cantonment areas across the country for the purpose of the fairly innocuous "Grow More Food" campain in the 1950s, provoked ire within the army. The proposal was soon dropped. Then again, the idea of a large-standing "people's liberation army" of about 10 million soldiers is unacceptable to most Indian politicians, who fear the loss of civilian control of the military and eventually to military coups and praetorian regimes.

The Military as a Disintegrating Force

The Indian military sees itself as the defender of the

nation and an important symbol of national unity. They would like to keep it that way. The military as a disintegrating force arises mainly from its reluctant role in the maintenance of internal security. Controversy over the internal use of force to quell armed insurgencies, of course, surrounds both the paramilitary forces and the armed services. In the case of paramilitary forces, the issue is not whether such forces are needed for the maintenance of internal security, but whether these forces are likely to be misused by the government in power. In the case of the armed services, the question is whether such forces should be used at all in the maintenance of internal security.

Internal Security. Almost to a man, the leaders of the armed forces are convinced that the military should not be used to maintain internal security. First, the military argues that to use the armed services "against their own people" would not only produce a breakdown in the military-civilian trust that has been built up over four decades, but, if deployed on internal security missions in the border provinces such as Punjab, would also undermine the army's ability to fight a war because of the alienation of the people living in those regions (Thomas, 1986). The military will be seen eventually as the enemy of the people, a situation that may be found in Pakistan and several other Third World countries in Asia, Africa and Latin America. In Pakistan, the military regime of Yahya Khan had a record of brutal suppression in Baluchistan and the former East Pakistani province which eventually broke away to become Bangladesh. The military regime of General Zia ul-Haq was engaged in political battles with various opposition groups in Pakistan, especially the faction led by Benazir Bhutto that had a large Sindhi following. This is precisely the kind of situation that the military in India would like to avoid.

If the deployment of the military as a constabulary force becomes unavoidable, two modus operandi agreed upon by both civilian and military authorities will have prevailed. One of them is the classical colonial use of "alien" troops to wage counter-insurgency warfare. Hence the Sikh Light Infantry and other lightly armed and mobile north Indian units (except for Gurkha or paramilitary units like the Assam Rifles who may feel some racial empathy with those involved) were fielded against the Chinese-backed guerrillas active in the 1970s in the northeastern tribal states of Nagaland and Mizoram. When this option is considered too risky for social or political reasons, the preferred solution is to deploy the integrated Brigade of Guards, so as to ensure that no stigma attaches to any one communal regiment as the executor of "police action."

Thus, ordered into action by a Sikh lieutenant general, and led by a Muslim Colonel, the Brigade of Guards elite strike force that included Sikh soldiers stormed the Golden Temple in Amritsar in 1984 and dislodged well-armed Sikh secessionists (Kapar, 1984). The selection of a Sikh commander to lead "Operation Blue Star" was not coincidental either, it being consistent with the colonial practice of letting the neutralization of a rebellious community be carried out under the leadership of persons from the same community. In the last of such actions during the British Raj, for example, a Muslim Deputy Commissioner in the 1930s under the instructions from the Muslim Chief Minister of pre-partition Punjab, Sir Sikander Hayyat Khan, led a contingent of mostly Muslim troops against armed Muslim fundamentalist dissenters holed up in a mosque in the Jhelum district of Punjab (now West Punjab in Pakistan), and put an end to the so-called Khaksar Rebellion.

Second, the armed services are not equipped or trained to deal with problems of internal law and order. As Lieutenant General M. L. Thapan noted: "A fundamental principle of war is concentration of men and materiel at the right place and at the right time." (Statesman, 3 August 1980). Internal security duties, on the other hand, "require dispersion and the use of minimum force since our own countrymen are involved." Lt. General A. M. Vohra observed that because the army does not mingle with the crowd as the police are required to, the army's ability to sense and deal with internal riots is severely limited.

Third, the use of the armed services for internal security may produce a breakdown in military training and readiness for dealing with external security. Their politicization and corruption may also be encouraged. Frequent use of the military for internal purposes would invariably arouse the emotions of both soldiers and civilians and could eventually result in struggles between the two sides leading to military coups and takeovers.

Despite military opposition, civilian leaders feel that situations may sometimes exist in which limited military participation becomes inevitable even if undesirable. Jagjivan Ram, former defense minister in the Congress government of Indira Gandhi and subsequently deputy prime minister in the Janata government of Morarji Desai, claimed that in the cases of extreme breakdown of law and order, the mere introduction of army units in troubled areas tend to produce a pacifying atmosphere without these units having to resort to any actual force (personal interviews). Indeed, the army rarely acts when deployed in strife-torn areas and this reduces the deaths usually caused when police are compelled to open fire on violent mobs. However, military leaders interviewed in 1981 pointed out that this was true only because the armed forces had thus far been used sparingly. Frequent use, on the other hand, could destroy the pacifying effect, and consequently they feel that the goal of the central government should be the total avoidance of the use of the armed services for internal security.

However, in two domestic areas military leaders agree that the use of the armed services may prove unavoidable. The first concerns the problem of dealing with insurgencies, as in the northeast of India. But even here the military caution against its frequent and continued use. Persistent insurgencies like those in Nagaland or Mizoram, suggest the need for political settlements rather than military solutions. It was advice taken to heart by the Government of India, which has negotiated deals with Nagas and Mizos whereby the rebel leaders have accepted Indian nationality, foresworn violence, entered the political mainstream, won elections, and are now holding power in their states. The second area where the military agree it has a contribution to make is in disaster relief. There is general consensus within the military that this constitutes one of their essential domestic tasks.

The Indian Army's misgivings over its use in the maintenance of internal security appears to have been justified. Although there have been several disturbances in the police and paramilitary forces, the army's assault on the Golden Temple in June 1984 set off the first, though brief and limited, military mutiny in independent India. Only two other mutinies had previously occurred in the history of the Indian armed forces, and both had occurred under the British. The first was the Great Indian Mutiny of 1857-58 when Indian sepoys (foot soldiers) revolted against their commanding British officers. The second was the mutiny by Indian seamen of the Royal Indian Navy against their British naval commanders in 1946. In retrospect, both mutinies were seen as milestones in the road towards Indian independence. So too the Sikh mutiny in the Indian Army is claimed by some Sikhs as a major step towards an independent state of Khalistan. An independent Khalistan would then herald the beginnings of the disintegration of India.

The Sikh mutiny of 1984 raises fundamental questions about the domestic role of the armed forces. The short term need after the army assault on the temple was to restore the holy shrine to Sikh civilian control and to pacify the anger felt by all Sikhs throughout India and abroad. From a military standpoint, it was also important to restore the confidence and commitment of the army's Sikh soldiers, who have historically distinguished themselves in battle through tremendous valor and sacrifice. Since the assassination of Mrs. Gandhi in October 1984 and the succession of Rajiv Gandhi as prime minister, these short term goals have been addressed with limited success.

In the long term, the solution may lie in separating (or shielding) the military from the role of maintaining internal security. In fact, representations to this effect made by the military authorities to Prime Minister Indira Gandhi, spurred the government's decision to raise specialist paramilitary forces like the Border Security Force, the Indo-Tibetan Border Police, the Industrial Security Force, and the Central Reserve Police Force (Thomas, 1986:72-77). This will require a clear delimitation of functions among the military, border security forces, and paramilitary forces. The military's role should be strictly in the realm of external security. (Note, however, that the concept of such a narrowly defined use of the military runs into formidable obstacles, primarily with the "Kitchenerian" colonial legacy embodied in the Indian constitution which legitimizes the military's "Aid to the Civil Authorities.")

Special counterinsurgency units may be deployed to deal

with guerrilla warfare, particularly where this involves external military assistance to the guerrillas. But even here the problem is basically internal and political, and calls for appropriate solution at these levels. Despite fears of the increasing growth and misuse of paramilitary forces, their use may be preferable to the induction of Indian army units into riot-torn areas. The creation in 1985 of the National Security Guards, drawn mainly from regulars or early retirees of the Indian Army but distinguished functionally and operated separately from the regular armed forces, may provide the well-trained and disciplined force needed to deal with severe acts of civilian violence within the country.

   Threats to Civilian Rule. Does the military pose a threat to civilian democratic rule? Will this ultimately lead to destabilization and the disintegration of the nation? The military thus far has not seized power in India and appears to show no interest or indication of doing so. This may seem remarkable given the size and power of the armed forces in India. Military leaders in uniform or in civilian clothes have ruled at various times in South Korea, Taiwan, Philippines, Indonesia, Thailand, Burma, Bangladesh and Pakistan, to name only the Asian states of the Third World.

   Apart from the declared lack of interest by the military in seizing political power in India, there are several difficulties in carrying out a military coup against civilian rule. As Lieutenant General Vohra described it, army organization into five regional commands makes it difficult to coordinate a strategy that would ensure that all regional commanders would cooperate in the military seizure (personal interviews). Then again, the ethnic diversity of the military in India might produce resistance by some groups who would fear "northern," Hindi-speaking, or Punjabia domination of the country. Even if there were no difficulties here, there would be no guarantee that the Indian Navy and the Indian Air Force would cooperate in the staging of a coup against civilian rule. As such, the army has argued that military coups are almost impossible to stage in India even if the army wanted it, which they claim they don't.

   The difficulties of staging a military coup in India arise partly from the fact that there is no equivalent of a Joint

Chiefs of Staff (JCS) in India (Thomas, 1986: 130-133). Instead, each of the services is commanded by a Chief of Staff. Although the Chiefs of Staff of the army, air force and navy meet regularly in the Chiefs of Staff Committee (CSC), this is merely a consultation body and not a consolidated executive body. Indeed, repeated calls by the army to establish a Chief of Defense Staff (CDS) organization equivalent to the CDS in Britain or the JCS in the United States have been rejected by the civilian authorities. Politicians fear that such a consolidated body directing all three armed services could pose a threat to civilian rule. This is so sensitive an issue with the political leadership that it allegedly cost Lieutenant General S. K. Sinha, the Deputy Chief of Army Staff and the most influential proponent of Combined Chiefs of Staff organization, the top post in the army. Lieutenant General Arun Vaidya superseded him in 1983 as next the Chief of Army Staff with the rank of full general.

There has also been some opposition from the navy and the air force to the idea of a CDS because they fear that the army may dominate the CDS to their disadvantage. The army, however, has countered that the position of Chief of Defense Staff could be rotated among the three services, as is the case in the Chiefs of Staff Committee, so as to prevent army domination.

It is probably not a mere coincidence that as the power of the military has grown in India, its inputs into the security policy-making process has declined. The growing number of manpower and firepower of the military has been offset by increasing civilian control of the military. At one time, defense decision-making involved three basic committees in a three-tier system that included a constellation of other advisory committees (Thomas, 1986:119-133). These were the Defense Committee of the Cabinet (DCC), the Defense Minister's Committee (DMC), and the Chiefs of Staff Committee. The Chiefs of Staff of the three armed services were formally a part of the DMC chaired by the Defense Minister, and of the CSC chaired on a rotational basis by each of the Chiefs of Staff. They were also informally invited to attended the DCC on matters concerning the military and defense issues.

In the mid 1970s, the DCC was expanded into the Cabinet

Committee on Political Affairs (CCPA) that was intended to deal with both external and internal security matters. The CCPA included virtually every cabinet minister in the central government since security problems were broadly envisaged to include problems of resources and effects on economic development, as well as the implementation of defense programs. But since the CCPA dealt with sensitive internal political issues as well, the Chiefs of Staff, for all practical purposes, were excluded from its deliberations.

Similarly, the DMC was replaced by a wider body known as the Defense Planning Committee (DPC). The DPC met under the chairmanship of the Cabinet Secretary and included the Secretary to the Prime Minister, the Secretaries of Defense, Defense Production, External Affairs, Finance and the Planning Commission, together with the three Chiefs of Staff. Since the DPC consists of seven high-ranking civil servants and the three service chiefs, the military has claimed that the preponderance of the civilians has worked against military inputs into defense policy-making. It was also argued that the new committee provided no direct interaction between the defense minister and the service chiefs, and that the committee might eventually erode the authority of the defense minister himself, since the DPC headed by the cabinet secretary could report directly to the PACC. This situation--tightening civilian control over an expanding military--may prove unsettling if the military becomes disgruntled with its eroding role in the defense decision-making process coupled with its increasing role in maintaining internal security.

The dominance of politicians and civil servants in the defense decision-making apparatus has led, the Indian military feels, to the absence of professionally devised strategy and of long-range military planning (personal interviews). The Strategic Planning Group headed by the Prime Minister and including the Defense, Finance, Home, and External Affairs ministers may have been established to address this aspect of the military's discontent. The Strategic Planning Group, an apex body, is expected to provide the long-range perspective for security to the year 2000 and will have direct inputs from another new body, the Defense Planning Group, chaired by the Defense Minister and having the Ministers of State for Defense and Defense Production, the Secretaries for Defense, Defense Production, and External Affairs, and the three Service Chiefs as members (Hindustan Times, 6 April 1987). It is not clear whether these newly constituted groups will replace the CCPA and the DPC, and whether they will prove to be the vehicles for more satisfactory participation by the uniformed services in Defense-related decision-making.

Conclusions and Prospects

On balance, the military makes a positive contribution to the development and stability of India. It provides good employment, training and discipline to a large cross-section of the Indian population who may be gainfully employed once they have finished their military service. As such, there are good economic reasons for the increasing competition throughout India to join the armed forces. In a country where there is a high level of unemployment and an even greater level of underemployment, the military is a very attractive source of employment, benefits and pensions. Even during the British Raj, there was never any need to introduce the draft. During both World Wars, millions of Indians volunteered, joined and served the British Indian Army in its overseas campaigns. This trend has continued in the post-independence era with more applicants every year than there are available spaces. The restrictions to entry are usually disqualifications based on physical fitness and military manpower ceilings.

In 1988, the uniformed personnel of the military in India was about 1.4 million strong: 1.2 million in the Indian Army, 115,000 in the Indian Air Force, and 52,000 in the Indian Navy. The military also employs another 2 million civilians to perform various "white" and "blue collar" services. These figures have only marginally changed over the last 25 years, in spite of the fact that the population has increased from about 440 million to about 800 million in 1987. In proportion to the size of the growing Indian population, the size of the military in India has relatively decreased significantly, thus reducing the job opportunities available to the growing Indian work force. This proportionate decline in the size of the military will increase the competition and the tensions among various communities in India who view the military as a gainful source of employment and security.

However, the major problem that threatens national unity and may involve the military is not the competition, recruitment and composition of the Indian armed forces. In recent years, there have been increasing stresses on the Indian political system as a whole that threaten its eventual stability and survival. When adverse political conditions reach a breaking point, the military may be expected to play a crucial and decisive role in the political outcomes that will eventually prevail. The belief that such a potential role does exist is both a source of assurance and worry for Indian politicians.

The role of the military as the underwriter of India's ultimate unity and stability would be unlike its earlier role during the prolonged political crisis in British India that eventually led to the partition of India and the creation of the new state of Pakistan. In the partition crisis of 1947, the Indian military remained apolitical and uninvolved, despite the emotional stress aroused by widespread communal killings among Hindus, Muslims and Sikhs. This is because the partition of the subcontinent left the new India under the control of some of the stalwarts of the nationalist movement-- Jawaharlal Nehru, Vallabhai Patel, Maulana Azad and Gobind Ballabh Pant. Their presence and unity appeared to guarantee stability and direction to the fledgling Indian Union, and reassured the military about the bonafides of their new civilian masters. They infused the hitherto mercernary colonial force with the nationalistic spirit and respect for the political leadership which had fought for independence.

This situation was in contrast to the inability of Pakistani politicians after Jinnah to secure respect from the Pakistani armed forces leading to the 1958 military coup staged by General Ayub Khan and the subsequent preeminence of the military in the Pakistani polity. Since then there have been brief periods of civilian control in Pakistan, first under Zulfikar Ali Bhutto from 1972 until his overthrow by General Zia-ul Haq, and then under his daughter, Benazir Bhutto, now also deposed. Benazir Bhutto's ability to gain control through national elections was aided accidentally by the unexpected death of Zia-ul Haq along with seven of Pakistan's eight four star generals in an unexplained plane crash in July 1988. The only general that did not board that plane was Mirza Aslam Beg, a Muslim originally from India, and who became the Chief of Staff of the Pakistan Army.

The problem in India today is the absence of any viable national leaders, at least apart from Rajiv Gandhi. The former Indian finance minister who also held the post of defense minister, Vishwanath Pratap Singh, later broke with the Congress party and may have the makings of a national leader in the future. But such possibilities have come and gone in the past without providing leadership succession. The Indian political system does not seem to generate leaders with widespread national suppport. There have been only two prime ministers in India that did not belong to the Nehru line of dynastic-type succession, and both tenures have been brief. Prime Minister Lal Bahadur Shastri did not live long enough after his strong leadership during the 1965 Indo-Pakistani war to achieve a national following. Prime Minister Morarji Desai of the 1977-79 Janata government was a compromise leader of the Janata party, the candidate himself having been elected by the people from a single constituency in Gujerat. His political following never went much beyond his own state. That the Nehru familiy has held prime ministerial power for three generations is testimony to the fact that India constantly faces a leadership crisis. There appears to be a potential leadership vacuum in India since it seems unlikely that Rajiv Gandhi's son will succeed him to office in the future. Without such national leaders who can appeal to all segments of the country, India will always face the threat of disintegration. When such adverse political circumstances prevail, the military may be perceived as the government of last resort that could bring about unity and stability to India.

References

Cohen, Stephen. 1971. The Indian Army: Its Contributions to the Development of a Nation. Berkeley, CA: University of California Press.

____ 1984.   The Pakistan Army.   Berkeley, CA: University of California Press.

Finer, S. F. 1976.   The Man on Horseback: The Role of the Military in Politics.   Baltimore, MD: Peregrine Penguin Books.

Kapar, Rajiv.   1984.   Sikh Separatism: The Politics of Faith.   Winchester, MA: Allen and Unwin.

Kravic, Lorne.   1967.   India's Quest for Security: Defense Politics 1945-1965.   Berkeley, CA: University of Califorina Press.

Lowenthal, Abraham, and John Fitch, eds. 1986.   Armies and Politics in Latin America, second edition. New York: Holmes and Meier.

Nordliner, Eric.   1977.   Soldiers in Politics: Military Coups and Governments.   Englewood Cliffs, NJ: Prentice-Hall.

Perlmutter, Amos.   1977.   The Military and Politics in Modern Times.   New Haven, CT: Yale University Press.

Simon, Sheldon, ed. 1978.   The Military and Security in the Third World: Domestic and International Impacts.   Boulder, CO: Westview Press.

Sinha, S. K. 1980.   Of Matters Military.   New Delhi: Vision Books.

Thomas, Raju G.C. 1986.   Indian Security Policy.   Princeton, NJ: Princeton University Press.

7

THE MILITARY AND SOCIAL INTEGRATION IN ETHIOPIA

Claude E. Welch, Jr.

Six facts must be borne in mind in this survey:

1.   Ethiopia has experienced probably the most sweeping social and political revolution on the entire African continent.

2.   Leadership of the revolution emerged from junior commissioned officers, who forcibly eliminated other rivals to power and, by the 10th anniversary of their 1974 ouster of Emperor Haile Selassie, formally supplemented their coercive control with a workers' party organized on classic Leninist lines.

3.   Simmering regional disputes pose continuing challenges to the central government, with its control over main roads and towns liable to pressure from groups based to a large extent on ethnic differences.   National integration, in other words, remains incomplete.

4.   The Ethiopian armed forces are, far and away, the largest in tropical Africa, with the impressment of peasants into a hastily raised militia in 1977 and with all males of draft age having been subject to compulsory service since early 1984.[1] Rebellion justified the further enlargement of what already had been considerable military power.

5.   Ethiopia moved, in less than a decade, from the leading recipient of American military aid south of the Sahara to the leading recipient of Soviet military aid in the area; this

assistance resulted from growing ideological affinity, overt problems of national unity, and Soviet calculation of significant geostrategic advantage.

6. Ethiopia remains among the poorest of states in the poorest of continents, with military costs weighing heavily on a government faced by periodic drought and famine as well as regional rebellions.

From the interplay of these factors comes the fascinating story of Ethiopia under Mengistu Haile Miriam, Chairman of the Provisional Military Administrative Council and the PMAC Standing Committee; Chairman of the Council of Ministers; Secretary General of the Workers' Party of Ethiopia. The country cannot be understood without examining the background, views and policies of Mengistu. These policies, in turn, must be considered in terms of Ethiopia's economic, ethnic, religious and social kaleidoscope, notably as these were manipulated for close to sixty years by the "Lion of Judah," Emperor Haile Selassie, against whose crumbling personal leadership the 1974 revolution erupted. Cleavages in Ethiopia, papered over under the ancien regime, were intensified with the profound popular mobilization of the mid-1970s. Regional and ethnic rebellions have both challenged and rationalized Mengistu's strong personal leadership, with the collapse of the imperial government necessitating a search for a different basis of ruling. The complex relationships between "class" and "nationality" have yet to be fully resolved in Ethiopia.

The governing officers' solutions to internal dissent - tactical initial cooperation with some civilian groups, followed by their subsequent repression and/or political cooptation; strong military pressures against most opponents; efforts to establish an ideological foundation in sharp contrast to the "traditional" legitimation sought by the deposed Emperor - raise fascinating questions. To what extent was the revolution a consequence of inherent limitations in Haile Selassie's style of rule? Did the training and outlook of PMAC leaders make them more likely to use force in the face of social cleavages than civilian leaders might have done? Among its many possible causes, was the revolution fueled primarily by latent class antagonisms, by awakening ethnic sentiments, by sharp conflicts within the relatively limited urbanized intelligentsia (including military officers), by external pressures and inducements, or by the strong will of an ambitious, power-oriented ruler? Can a country in which more than 80 nationalities have been identified, but in which only one has exercised substantial political power, achieve "true" integration that is perceived as treating all groups fairly? Does compulsory military service enhance national unity, in the face of sharp regional antagonisms? How can a party allegedly organized on the basis of workers, but in fact dominated by the younger officers who have governed since late 1974 - serve as a "vanguard" for revolutionary transformation? Finally, to what extent have the profound alterations since 1974, and their consequences for the integration - or disintegration - of Ethiopian society, resulted from a constellation of factors peculiar to Ethiopia itself, rather than from factors general to the developing world?

In keeping with other chapters in this book, this section examines national socio-cultural cleavages, their reflection in the armed forces, the effects of military service on socialization, efforts at intra-military social integration, and the political and military consequences of current policies. The existing Ethiopian context does not make research easy, however. Diplomatic relations with the United States were formally broken in July 1980, having dramatically worsened as the revolution swept away the persons with whom the US had been accustomed to deal;[2] the unsettled conditions of much of the country lead to unverifiable claims by both the regime and its many opponents; the internal workings of the armed forces, and in particular the nature and extent of political awareness among drafted peasants, can only be inferred rather than investigated directly. Hence, the pages that follow rely far more on informed journalists' observations than on direct field research, on educated surmises about internal conditions more than quantitative, verifiable or scientific findings.

Ethnicity and Regionalism in Ethiopia

Cultural pluralism characterizes Ethiopia. The country's first major census was taken in May 1984, covering an estimated 85 percent of the population. It showed significantly more people than anticipated: in contrast to the official figure of 33,680,000 in mid-1983, the census (including estimates for areas not reached by enumerators) found 42 million inhabitants. (Africa South of the Sahara 1986: 416.) Ethnic affiliation was not available -- and, given the ideological proclivities of the government, may not have been asked. The diversity is significant, however, since culture interacts with geography. Four major language groups are represented, with two dominant: Ethio-Semitic, including nearly half the total population; Cushitic, with close to 40 percent; Omotic; and Nilo-Saharan. No less than 82 mother tongues are listed, with number of speakers ranging from 250 to close to 8 million (Legum 1982-1983: B156-7). A leading scholar of Ethiopian culture (Levine 1974: 33-39) has classified the population into nine groups: North Eritrean, Agew, Amhara-Tigrean, Core Islamic, Galla, Lacustrine, Omotic, Sudanic and Caste groups. But far and away most important, three major groups account for three-quarters of the populations: Oromo, 40 percent; Amhara 19 percent; Tigrean 16 percent (Schwab 1985: xv). The ethnic and linguistic diversity is counteracted by a historic cultural core. The Amhara heart of Ethiopia -- what historians called Abyssinia -- served as the focus from which 19th century emperors expanded their control. In Levine's judgment, Ethiopia is "the only country in Africa with a large number of ethnic groups where one of these groups has imposed its rule and its language over the rest and has preserved indigenous national institutions, elites, and culture patterns from displacement by Western forms and authorities" (Levine 1965: 3). To put the matter more bluntly, the growth of Amhara territorial control under Menelik II, its consolidation under Haile Selassie, and continued political centrality of Amhara under the revolutionary government in combination created a polyglot state in which one nationality -- by no means the largest -- achieved and continues to maintain dominance.

Amhara historians view the 19th century extension of imperial control as reinstating what had been challenged by the earlier movement of Oromo (Galla) and other peoples into areas of traditional Amhara dominance. The Oromo migration, starting in the early 16th century and extending over many decades into regions vaguely subject to the Christian crown, dramatically altered a relatively cohesive, confined social and political system. In Markakis' judgment, "Ethiopian tradition neither forgot nor forgave the Galla invasion and the loss of land occupied by the invader... Galla territories never ceased to be regarded as rightfully belonging to the domain of the Solomonic throne" (Markakis 1974: 23). The Amhara had been both pushed out from, and outnumbered in, "their own" country, as traditionally defined. The situation seemed to invite response. And, while Europeans were carving up other parts of Africa, Ethiopian rulers were conquering -- or in their view regaining -- territories lost to external invaders. Menelik's reign (1889-1913):

> "turned Ethiopia into an empire in fact as well as name. In a burst of furious energy, the Ethiopians overran the southern part of the plateau, doubling the domain of the Solomonic throne and imposing its rule on a large number of peoples of diverse origin and cultures... The homogeneity achieved in the northern plateau through a centuries-long process of integration within the framework of the Christian state was now overshadowed by the incorporation of numerous, sizeable alien groups" (Markakis 1974: 22-3).

By far the most sizable of these groups is the Oromo, whose role in Ethiopian history has never been proportionate to their numbers. This political weakness has been explained through the nature of their spread and their partial assimilation into Amhara society. The Oromo lack political unity; they are a confederation of tribes, never politically united, who expanded dramatically in the 16th century[3] into areas the Amhara themselves had colonized in the previous century (Baxter 1978: 284). Some Oromo adopted Islam, others Christianity; political structures evolved in different directions, ranging from the decentralized and acephalous to the complex and unified. Common to all Oromo values, however, have been age- and generation-grading. Office holders are bound within "a combined ritual, political and military organization which maintained a strongly democratic and egalitarian ethos and restrained the exploitation of office, wealth and power." (Ibid.: 284-5).

Such values, egalitarian rather than hierarchical; solidaristic rather than individualistic; fused rather than separated religious and political functions -- stand in sharp contrast to Amhara values (Levine 1974: 128). For several centuries, the Amhara have counteracted endemic regionalism with a sense of cultural and political unity. References to a "national" community, to a monarch standing above the various princes and military commanders, and to the Coptic Church, distinguish the Amhara from the Oromo. The latter, one specialist argues, lack the essential underpinnings of unity. "If by nation one means a sizable group of people who have some sense of belonging to a single societal community by virtue of sharing important past experiences and a common historic destiny, then the Galla [Oromo] do not constitute a nation, nor have they since their appearance as significant actors in the arena of Greater Ethiopia during the sixteenth century" (Ibid.: 135). The Amhara, by contrast, have enjoyed a history, institutions, values and fears that jelled into a profound sense of their own singularity and right to lead. As one critic has written,

> Their preservation of a subtle, literary and Christian cultural tradition in beleaguered isolation in the Ethiopian Highlands is an achievement in which the Amhara take a very proper, if exaggerated, pride and one which one should not denigrate. The ruling elite never seems to have doubted the absolute superiority of its own culture and its duty to impose it on any who sought near equality with it. But, since the sixteenth century, fears of Islam and of the Oromo have dominated the political consciousness of the Amhara ruling elite, and the thought of the two in combination has been their recurring nightmare (Baxter 1978: 285).

Cleavages within pre-revolutionary Ethiopian society followed economic as well as cultural lines. Class divisions prior to 1974 interacted with ethnicity. The pre-1974 role of class in Ethiopia has (as might be expected) aroused controversy among specialists. Clearly, class conflict existed in embryo, notably in the south. The emergence, in Markakis' phrase, of a "vast class of landless peasants" meant that "the role of ethnicity [in Ethiopia] is conditioned by, and usually subordinated to, class considerations" (Markakis 1974; 6-7). On the other hand, since the majority of these peasants were Oromo, and the overwhelming majority of landowners were Amhara, economic and ethnic factors were inextricably mingled.

Collapse of the traditional Amhara landed-class dominance came with the deposition of Emperor Haile Selassie in September 1974, the result of a "creeping coup." The discontents of the military rank and file and NCOs provided much of the fuel. Ouster of the ancien regime came under the shadowy leadership of the "Derg", a 120-member coordinating committee composed of three representatives from each of the army's 36 battalions, plus representatives from the police, other armed services, and the military schools. Battalion officers and other higher officers were automatically excluded; hence, over half the delegates were NCOs and privates. To understand better the rise of the Derg, we must turn to the transformation of the Ethiopian military in preceding years.

The Military Experience and Socialization

Ethiopia rightfully claims one of the longest, proudest martial traditions in Africa. The survival of the Abyssinian heartland over many centuries, its expansion while other African societies were being subordinated during the "Scramble" for Africa, and its resistance to successful European invasion until 1935 are unquestionable historical facts. Military virtues stood high in traditional social values. The skills and bravado of the individual warrior were stressed, with battle both recognition of and reward for personal heroism. The successful soldier earned high status. Such a heritage inevitably plays some part, even in the revolutionary Ethiopia of the late 1980s.

In addition to the historical cultural continuity, however, new elements of military socialization have been introduced in the past 50 years. The introduction of modern officer training through exogamous training on a modest basis just prior to the Italian conquest, and its significant expansion with primarily American assistance after World War II, emphasized discipline, hierarchy, and use of contemporary military equipment (including airpower). New concepts emerged: "professionalism," with its concomittants of centralization, expertise and corporateness (Abrahamsson 1972: 15; Huntington 1957: 8-10); "nationalism," with a sense of Ethiopian society as a whole, in its multiethnic character; "modernization," with the armed forces as a potential model for the revision of narrow or outdated loyalties.   The sweeping revolution of the mid-1970s thrust segments of the armed forces into the center of the political stage.   The quest for ideological legitimation, and in particular the groping toward the Workers' Party, stressed Marxist-Leninist elements.   Foreign advisers, materiel and training - now from Eastern rather than Western sources - inevitably influenced military outlooks.   And, probably most important, the quelling of regional separatist movements gave the armed forces direct combat experience on a scale and level unmatched in sub-Saharan Africa (with the only possible exception being Nigeria during its 1967-70 civil war).

The pages that follow treat military socialization on three levels, during three periods: for the rank and file, for non-commissioned officers, and for commissioned officers; prior to World War II, prior to the Revolution, and since 1974. Sharp contrasts exist among levels and periods - but also marked continuities.

The "Traditional" Setting

Near-continuous warfare marked the historical Ethiopian/ Abyssianian state. Much of the combat was internal: the strife of regional leaders (a ras or negus against each other), as provinces vied for supremacy. But an even greater part came into protection against external invaders, particularly with the irruption of Islam into the highlands (Trimingham 1962). Martial skills and values loomed large in Amhara and Oromo traditional values. Military courage was prized in both, though for different reasons.   Among the Amhara, prowess in battle was rewarded through booty; among the Oromo, valor was proof of an individual's emerging manhood.   Bravery in warfare, in one specialist's judgment, had mixed results for the latter: "it was their democratic collectivism and their historical project, the very factors which made the Oromo unsuited for extending political dominion over others, that gave them a special short-term advantage over the Amhara in waging aggressive warfare" (Levine 1974: 155). According to the same author, six factors produced an "effective soldiery" among the Amhara and Tigreans: physical capacities; use of arms from childhood; a cult of masculinity conceived in terms of military prowess; the prospect of acquiring booty; success in military activity leading to social mobility; and vertical loyalties to one's leader (Levine 1968: 7-8).

In traditional society, combat was the responsibility of all adult males -- though not all men were always engaged in combat. Leadership was vested in the nobility -- a noblesse de l'epee, since titles were, "first and foremost, military titles" (Levine 1965: 158). (For example, dejazmatch refers to "general of the gate," responsible for the center of the formation.) No permanent, standing army (other than armed personal retainers of local lords) existed in pre-World War II Ethiopia; forces were levied for individual campaigns on ad hoc bases. Soldiers were expected to live off the land and provide their own weapons. Professional training per se did not exist. Leadership and loyalty were intensely personal. Rewards, particularly land, were distributed to the victors. The authors of the area handbook aptly summarize the heritage:

> Soldiering has been regarded as the surest path to social advancement and economic reward. Land, titles, and political appointments were traditionally awarded to those who proved their loyalty and competence in battle. As a result, a warrior's loyalties were directed toward a strong man, who could best assure his followers the fruits of victory, rather than to an abstract notion of the state or to government authority... Once it was engaged, the army's basic fighting unit was the individual soldier, and battles were usually decided by a final charge to bring the enemy to hand-to-hand combat... Leadership was intensely personal... Entirely predatory in nature, the army lived off the ruler's subjects wherever it camped in his domain (Nelson and Kaplan 1981: 240-2).

Such an organization permitted the extraordinary expansion of Ethiopian - or, more accurately, Amhara - control under

Menelik (Marcus 1975), and surprised Italian invaders in 1896. Based essentially on parochial loyalties (that is, personalistic and provincial), these armed hordes offered a shaky basis for strong central political leadership.   No collective training procedures existed; "the existence of a ready supply of capable soldiers enabled the traditional military system to function on a highly individualistic basis" (Levine 1968: 9).   Provincial autonomy was a far more salient goal than "national" unity in the "traditional" setting, while the concept of a multi-ethnic Ethiopia with roughly proportional representation of all groups in senior posts was utterly unthinkable. The survival of Abyssinia had been predicated on its Amhara, Christian character; its expansion of control over polyglot Ethiopia involved cooptation and Amharization of leaders from other groups; its protection against external invaders rested on patterns that proved adequate until the massive Italian invasion of 1935.

From 1941 to 1974

Modernization and professionalism were introduced into the Ethiopian military on a large-scale basis after World II, although Haile Selassie's desire to stabilize, strengthen and extend central authority led him to some earlier steps.   The Imperial Bodyguard was established in 1917, with a handful of White Russian instructors; a permanent Belgian military mission provided for its training starting in the late 1920s; some Ethiopian officers were trained at the French military academy at St.-Cyr during this period; in 1934, Swedish advisers began to staff the school at Holeta.   Faced with persistent provincial loyalties and the power of individual leaders to maintain autonomy, Haile Selassie could claim but could not exert full control over the nobility.   The Italian conquest - in which the regular army and many of the provincial militias were wiped out, and as a result of which most of the small Ethiopian intelligentsia   was   executed   -   paradoxically   offered   the opportunity for rebuilding the center and weakening the periphery.   The expulsion of the Italians in 1941 was followed immediately by an agreement with Great Britain on military training. In less than a decade, the military power of the provincial nobility had been broken and forces loyal to the Emperor recruited and deployed. A standing army directly controlled by Haile Selassie "rendered the existence of separate regional armies unnecessary and possessed a capability clearly superior to that of any force a rebellious lord might muster... The traditional function of the noble as a military commander was thus made obsolete" (Levine 1965: 179). The calibre of training and the quality of equipment in the imperial forces were far above any earlier in the country. Modernization, in short, entered the Ethiopian military in fortuitous post-war circumstances. Haile Selassie could centralize and innovate with little fear of successful provincial rebellion. And, with the influx of major American assistance after 1952,[4] modernization and professionalization of the armed forces proceeded rapidly.

Military officers in the late imperial period had to display two conflicting qualities: loyalty to the emperor; and competence in command. Loyalty was sought through careful recruitment of those perceived as supporting the Solomonic dynasty; competence was achieved and proved through academy training and combat experience. Those chosen for advanced education were by no stretch of the imagination drawn proportionately from the major population groups. Amhara Christians were far and away the most likely choices for commissions; Tigreans and Christian Oromos lagged far behind; Muslims were "conspicuously absent" in the military hierarchy (Markakis 1974: 255). Estimates of ethnic composition vary: Lefort (1983: 18) asserts that in 1970 no less than 70 percent of the officers were Amhara, 10 percent Tigrean and Eritrean, and most of the remainder Oromo; Gilkes (1974: 247-8) says senior ranks were "disproportionately" Amhara and Tigrean, although the Oromo in the early 1970s comprised 21 percent of the Lieutenant Colonels, 30 percent of the lower officer ranks, and 40 percent of the rank and file.

At all levels, the officer corps was atomized. Writing in the early 1970s, Markakis commented,

Senior Ethiopian officers share the predominantly traditional, conservative social and political orientation of their counterparts in the governmental hierarchy. The profession of arms has not secured them greater initiative or independence from the autocrat than is granted to his other retainers. Rather it has brought them strict surveillance and more blandishments by the cunning monarch. The junior ranks are divided...by differences of social origin, education, and professional training, represented by the recruitment standards and training of the two military training establishments (Markakis 1974: 258).

The two schools referred to are Holeta military school (established in 1933) and the Harar military academy (established in 1957). They drew from different social strata – a fact with immense significance for the course of the revolution. Holeta admitted students with only elementary school education, and trained promising NCOs whose status may have been low; Harar in its early years conscripted men "from among the best secondary-school graduates and freshmen of the University College" (Markakis 1974: 253). Thus, the multiple divisions within the officer corps tended to fall into two broad categories, Holeta graduates and Harar graduates.

The Revolutionary Period

The revolutionary Derg has been dominated by Holeta alumni -- and, in particular, by Mengistu. Almost all Harar graduates in the armed forces were eliminated by 1977. In the blood-letting, senior officers were the first to go; they were followed by officers perceived as soft-liners on Eritrea (more attention will be given to the would-be Eritrean self-determination in the following section).

The Oromo were, as Halliday and Molyneux comment, originally perceived as the chief beneficiaries of the revolution, since they constituted the bulk of the southern plateau peasantry benefitting from land reform (Halliday & Molyneux 1981: 196). Recognition of Muslim as well as Christian holidays, and the use of Orominya in official broadcasts, suggested the Derg's willingness to broaden the cultural parameters of Ethiopian society. Since revolutionary leadership was vested in the military, and since the armed forces were predominantly Oromo,[5] it was widely presumed that the ouster of Haile Selassie stemmed in large measure from Oromo desires. Rumors spread as well about the purported non-Amhara origins of Mengistu Haile Mariam, based on his childhood residence in the Oromo-dominated province of Wollamo; in fact, however, he was probably born to an Amhara soldier and a mother from the traditional Black slave caste (Ibid.: 116).

Accordingly, although Oromo concerns have not been totally ignored since the revolution, Amhara primacy in the central government continues. In mid-1980, the Ethiopian Herald reported that two-thirds of the COPWE central committee were Amhara, leading one well-informed observer to comment, "It is clear that the Amhara are once represented in central officialdom far out of proportion to their numbers in society as a whole" (Schwab 1985: 54-5). I shall return to COPWE (the Commission to Organize a Workers' Party) later in this chapter; at this point, it is necessary to turn to the regime's ideological foundations and to the challenges it faces from regional rebellions.

Class, Ethnicity, and Ideology

The search for ideological justification led the Derg to explore many avenues. Its initial formulation came with the deposition of the Emperor, appealing to the people to reject the policies and personnel of the ancien regime:

"Therefore, people of Ethiopia, you must realize that the money which the King has snatched from the country and taken to a foreign house cannot come back to Ethiopia to help the hungry, to help the Ethiopia which is shedding tears, to help Ethiopia in her economic difficulties. Therefore, people of Ethiopia, you must know that you must now stand on your own feet and build your country through your sweat and through your blood if necessary" (Kirk-Greene 1981: 46).

This statement of September 11, 1974 was followed by a proclamation a day later, in which the term "Ethiopia First" (Ethiopia Tikdem) appeared. This proclamation declared Haile

Selassie deposed, the constitution suspended, parliament dissolved "until the people elect through truly democratic processes their genuine representatives dedicated to serve the interests of the people," and full power assumed by the armed forces (Text in Ibid 1981: 46-8). Optimistically, a statement on domestic policy proclaimed "abolition of ethnic, religious and age divisions..." Finally, a document dated December 20, 1974, set forth the origins and future directions of the movement of Ethiopia Tikdem. Among its five principles were two directly relevant to the theme of this book:

1. All Ethiopians of whatever religion, language, sex or local affinity shall live together in equality, fraternity, harmony and unity under the umbrella of their country. Ethiopia will become a country in which justice, equality, and freedom will prevail...

2. The rights of self-administration which our people had exercised at the village, district and regional levels and which had been usurped will be restored. The central government will be responsible for national or otherwise fundamental matters of state and give assistance and support to communities exercising self-administration...

In short: "Ethiopia Tiken means Hibrettesebawinet (Ethiopian Socialism); and Hibrettesebawinet means equality; self-reliance; the dignity of labour; the supremacy of the common good; and the indivisibility of Ethiopian unity." (Italics added) (Kirk-Greene 1981: 52).

This declaration was quickly followed by a series of reforms, decrees, and nationalizations of major sectors of the economy. A Provisional Military Administrative Council (PMAC) was created, which initially acted as an executive committee to the Derg and soon exercised full power. Proclamation of a republic in March 1975 and removal of rural lands from the Coptic Church showed the PMAC's determination both to destroy the foundations on which the ancien regime had been based, and to grope toward their own statement of principles. In increasing collaboration with MEISON (All-Ethiopian Socialist Movement), dominant officers tried to accommodate their concerns about class, ethnicity and regionalism within a statement of ideological principles.

The "nationalities" policy of the PMAC emerged in 1976 under the pressure of growing dissidence and regional pressure. The most important public pronouncements came in the April publication of the Programme of the National Democratic Revolution (NDR) and in the May Nine Point Statement on Eritrea. A few quotations from each will provide the flavor. First, from the NDR (published in full in Ottaway and Ottaway 1978: 211-16):

> 5. The right to self-determination of all nationalities will be recognized and fully respected. No nationality will dominate another one, since the history, culture, language and religion of each nationality will have equal recognition in accordance with the spirit of socialism. The unity of Ethiopia's nationalities will be based on their common struggle against feudalism, imperialism, bureaucratic capitalism and all reactionary forces. This united struggle is based on the desire to construct a new life and a new society based on equality, brotherhood and mutual respect...
>
> Given Ethiopia's situation, the problem of nationalities can be resolved if each nationality is accorded full right to self-government. This means that each nationality will have regional autonomy to decide on matters concerning its internal affairs. Within its environs, it has the right to determine the contents of its political, economic and social life, use its own language and elect its own leaders and administrators to head its internal organs (Ottaway and Ottaway 1978: 214).

Second, the "Policy Declaration of the Provisional Military Government to Solve the Problem in the Administrative Region of Eritrea in a Peaceful Way":

> ...[the people of Eritrea will enjoy] full participation in the political, economic and social life of the country... the government will at an appropriate time present to the people the structure of the regions that can exist in the future... [Discussions will be carried out with] progressive groups and organizations in Eritrea which are not in collusion with feudalists, reactionary forces in the neighbourhood and imperialists.

Note the caveats: any negotiations between the central government and Eritreans would be intended to "promote the unity of the oppressed classes of Ethiopia" (Ottaway and Ottaway 1978: 158).

As should now be evident, both documents were riddled with contradictions: they proclaimed the right to self-determination, while denying the opportunity for secession; they spoke of conflict within Ethiopian society, but believed this was expressed through class rather than ethnicity; they held out the olive branch of autonomy and compromise, while they made preparations for repression. By viewing the chief divisions within country as economically based, the PMAC misconstrued the nature of conflict. Its members believed the problem of Eritrea was a legacy of the imperial government with its divide and rule policies and its encouragement of class differentiation. The PMAC both underestimated the cultural roots of internal violence and was ripped by internal dissension in which the hard-liners, led by Mengistu, gained the upper hand. Hence, it is appropriate to turn to the rebellions the NDR and Nine-Point Statement were supposed to help resolve.

Regional and Ethnic Rebellions

Ethiopia has the dubious distinction of Africa's lengthiest guerrilla war. The armed resistance in Eritrea, initiated shortly after the former Italian colony was fully integrated into Ethiopia in 1962, in turn helped spawn similar movements for regional autonomy in the heterogeneous state. The pace of change stirred political and social awareness among all groups. The Derg moved into power as imperial authority collapsed, and with it was swept away the veil of Amhara dominance. The revolution unfolded in a setting of rapidly weakening central power, which Mengistu and his colleagues attempted to restitch together. By the late 1980s, some of the challenges had been militarily subdued, but continued to fester beneath the surface; others had been exacerbated by the regime's clumsy coercive policies. Social integration in Ethiopia may in fact have been hindered more than helped by the actions of the dominant military.

Space precludes any extended discussion of the Eritrean imbroglio, well treated in several publications (Erlich, 1983; Halliday & Molyneux, 1981; Legum and Firebrace, 1983; Selassie, 1980; Sherman, 1980). Suffice it to say that, after lengthy and complex debate, the United Nations General Assembly had voted in December 1950 to federate Eritrea with Ethiopia under the crown of Haile Selassie. The union thus restored the ancient center of Ethiopian culture, which Italian invaders had wrested away in 1896, gave Ethiopia its own ports, and added a center of education and industry to the relatively backward empire. The 1958 founding of the Eritrean Liberation Front (ELF) preceded by four years the full incorporation of Eritrea into Ethiopia, amidst allegations of widespread bribery of the regional assembly that voted for full merger. The launching of guerrilla struggle initially appeared to be a modest threat to the imperial government, although successful aircraft hijackings by the ELF attracted international attention. The strength of the secessionists appeared to increase as the power of the imperial government ebbed away -- although, as Erlich demonstrates, Ethiopia and Eritrea went through simultaneous revolutions that moved them in different directions (Erlich 1983). The bloody 1972 split between the ELF and the EPLF (Eritrean People's Liberation Front) could not mask a deteriorating security situation in the province. The fall of the emperor in 1974 brought into question all the earlier arrangements. As the Derg consolidated power in Addis Ababa and struggled with the issue of nationalities vs. class, ELF and EPLF forces gained control of almost all of the Eritrean countryside and most of the towns.

The Mengistu regime and the rebel forces remain irreconcilable in their basic objectives: the maximum the central government will permit is regional autonomy, within the framework of Ethiopia; the various Eritrean groups favor self-determination and independence. The lack of unity among the Eritreans has negated their chances for success; heavy handed actions by the Ethiopian government have precluded a mutually satisfactory resolution. Increasing reliance on Marxist ideology, clash of personalities, and an unwarranted assumption they would succeed with relatively limited guerrilla forces affected Eritrean resistance. Efforts to find a middle ground among the contending groups have foundered on conflicting aims, organizational rivalries and external influences. Pressures from major supporters of Eritrean movements (e.g., Saudi Arabia, which has contributed financially; Sudan, which has provided staging and resettlement areas on its own territory) have, as of this writing, failed to unite the factions. Ethiopia's key allies (the Soviet Union; Cuba) made possible the expulsion of Somali irredentists; however, they have privately questioned the wisdom of harsh Ethiopian policy toward Eritrea, having been in the embarrassing position of supporting Eritrean self-determination before the 1977 Soviet volte-face. Despite their disunity, the EPFL, ELF and others in the alphabet soup of Eritrean resistance continue to tie down large numbers of Ethiopian troops, to control significant parts of the rural areas, and to raise fundamental questions about the nature of self-determination in Africa.[6]

Although Eritrean nationalists controlled almost all the territory of the former Italian colony early in 1978, by the end of the year they had been pushed back into isolated pockets. The reason: overwhelming Ethiopian military might, made possible by unprecedently large assistance from the USSR. Massive mobilization of troops -- essentially a peasant army -- and close to $2 billion of Soviet equipment, transferred initially to oust Somalis from the Ogaden, made the difference. The hard line of Mengistu, embodied in the slogan "Unity [of Mother Ethiopia] or Death," triumphed over more conciliatory approaches suggested by his rivals. The integration of Eritrea is now being assured by force of arms. Military victory has been won, although the roots of resentment have been exacerbated rather than extirpated by Ethiopian actions.

The significance of continued Eritrean resistance lies outside the former Italian colony. As Legum pointed out, "Eritrea is first and foremost a test of the present Ethiopian rulers' intentions over national unity and reconciliation between the old empire's half-dozen major national groups and so, too, of its ability to evolve a national polity to fill the vacuum left by the Emperor. The fierce expression of Eritrean ethnocentricism [sic] stimulated similar nationalistic movements among the Tigreans, the Afars, the Oromos, Somalis and others -- all of whom came to demand either some form of regional autonomy or complete independence" (Legum 1984: B134). Faced with demands so basic about the fundamental framework of the state, the Ethiopian government has reinforced its search for class-based explanations of prevailing dissidence. Fundamentally, governing officers do not accept any claims to regional power that potentially threaten central control.

The Tigrean People's Liberation Front, claiming in a November 1983 memorandum to the UN General Assembly to control 85 percent of Tigray Province, has espoused a seemingly more moderate objective: "[the TPLF is fighting for] the national self-determination of the people of Tigray and for basic social changes. The TPLF is not a secessionist movement for it is not against the voluntary unity of the Tigrean people with others in Ethiopia" (Legum 1984: B135). With the racking problems of the drought, and the focus of international attention on famine relief in Tigray, Wollo and adjacent areas, the central government's problem could not be concealed.

With respect to the Ogaden area, dominated by Somalis, Ethiopian forces by late 1986 were in full military control, with occasional raids into Somali national territory. Cuban, Soviet and other advisers continued to help safeguard Addis's sovereignty over the arid eastern third of the country. The size of the Ethiopian army makes any successful internal challenge from the WSFL (Western Somali Liberation Front) or external challenge from Somalia unlikely.

Faced with endemic internal rebellion and intermittent external pressure, Ethiopian leaders have developed far and away the largest standing forces in sub-Saharan Africa.

The army accounts for 300,000 of the estimated 306,000 military personnel; the navy and airforce include 2,500 and 3,500 respectively. With more than 1000 tanks - far and away the highest number on the continent outside North Africa - plus 14 bombers, 90 transport aircraft, and a variety of vessels, the Ethiopian armed forces overshadow their neighbors. The army increased nearly 50% between 1976 and 1978, and then further quadrupled in size, rising from six divisions to 20 (Baynham and Snailham, 1983: 179). This rapid expansion, however, came at the cost of efficiency. Rising tensions based on ethnicity - notably between the Amhara and Oromo - have

Military Expenditures and Gross National Product, 1975-1985

Year	Military Expenditures Current Constant		Armed Forces thou.	Gross Nat. Product Current Constant $mil	$1982	Central Govt Exp $mil	ME/ GNP %	ME/ CGE %
1975	79	144	50	2077	3773	710	3.8	20.2
1976	111	190	65	2280	3895	766	4.9	24.7
1977	142	228	225	2499	4002	779	5.7	29.2
1978	127	190	233	2679	3996	923	4.7	20.6
1979	268	368	250	3087	4233	986	8.7	37.3
1980	339	425	250	3524	4429	1119	9.6	38.0
	/ Estimate \							
1981	352	404	240	3944	4522	1185	8.9	34.1
1982	376	406	240	4259	4589	1317	8.8	30.8
1983	390	405	240	4633	4805	1721	8.4	23.5
1984	420	420	190	4742	4742	1460	8.9	28.8
1985	411	398	240	4532	4392	1504	9.1	26.5

Source: U.S. Arms Control and Disarmament Agency,
World Military Expenditures 1985 (Washington: U.S. Government Printing
Office, 1985), p. 61; figures for 1983 military expenditures and central
government expenditures from Africa South of the Sahara 1986

(London: Europa Publications, 1986), p. 419, calculated at official exchange
rate of $1 = 2.070 birr. Constant military expenditures and GNP calculated on
1982 base. All expenditures in millions of dollars.

also been reported (Baynham and Snailham, 1983: 178).

Overall supervision of the armed forces has been vested since April 1984 in the National Defense and Security Council; it was established to help ensure more effective supervision of the government's conscription policy. Observers have noted negative consequences of the 30 months' compulsory service. For example, thousands of youth in Tigray, Wollo and Gondar provinces reportedly volunteered for the opposition Tigray People's Liberation Front rather than be impressed into the regular army (Legum 1984: B132). Increasing warnings from the government have to date failed to stem this hemorrhage. Indeed, given the large population and limited employment opportunities, the partial reliance on a draft, rather than full reliance on volunteers, appears counter-productive. Its quest for national unity may have led the Derg to underestimate the popular costs of conscription relative to the gains. The attempt to achieve a truly national army in the face of serious communal and regional divisions has been imperfectly achieved, at best.

Evidence of common military training leading to secular, national attitudes is not available. On the other hand, the rigor and discipline of armed forces' life appear to have precluded (or at least impeded) overt division. To the extent that dominant attitudes can be discerned, they appear to be based more on class and ethnicity than on pan-ethnic bases. Each deserves brief treatment.

A deeply seated common hostility toward the ancien regime linked many members of the armed forces with oppressed members of Ethiopian society prior to the revolution. The support of many Oromo for the Derg reflected the fact that the deposition of Haile Selassie and the initial reforms of the revolutionary government significantly improved the lot of peasants, and reflected as well the concerns of the military rank and file. In Lefort's flamboyant prose, the role of Mengistu Haile Mariam was paramount:

If his brothers-in-arms found him and then voted overwhelmingly for him, it was because he was perhaps the only one to know what he wanted, to say it, and believe in it firmly... In July 1974, when the army was faced with the question of power (but still more the question of power to do what?), he acquired his ascendancy because, in an army that was falling apart, he set the rebellious troops a task to carry out.

It was a task inspired by the deep intensity of the humiliations and frustrations that had enraged the other ranks. It was not at all by chance that, within the Derg, virtually all of Mengistu's supporters should be NCOs and other ranks... [Mengistu] was consumed with a class hatred, visceral and elementary, because it was mediated by no political or ideological apparatus. It was the same hatred, in its intensity and its brutality, as was felt by the peasantry in the south toward their landlords or the unemployed workers in Addis Ababa before the bourgeoisie. Therein lies the greatest common denominator between the 'masses' and Mengistu. (Lefort 1983: 277).

This outlook toward Mengistu has understandably softened with

time.   Ethiopia in the mid to late 1980s has become an
undeniable personalized dictatorship. The supreme leader's
power rests primarily on the military, secondarily on the
Workers' Party and earlier established peasant associations.

Oromo unity, feared by decision-makers in Addis Ababa,
has not materialized. Threats from the Somali appear minimal,
at least at time of writing.   The Amhara presence in the
armed forces is dominant, but not necessarily overwhelming.
Adherence to official ideology appears to help in promotion of
officers; yet personal loyalty to Mengistu and those around him
may be even more significant.

The combination of regional rebellion with drought and
famine has attracted international attention.   The costly
celebration of the 10th anniversary of the revolution - marked
by an ostentacious display of military hardware in Addis
Ababa while the ravages of starvation in areas north of the
capital evoked callous responses from decision-makers -
negatively affected the regime's external image. Governmental
restrictions on externally funded and staffed relief efforts in
dissident areas resulted in significant criticism.   The security
needs of the regime overrode citizens' access to food - and
common   wisdom   is   that   relief   supplies   reached
government-controlled areas far more rapidly and copiously
than they did rebel-controlled areas.   To some extent,
accordingly, the incredible human suffering in drought-parched
areas (notably Tigray and Wollo) was turned to short-term
political gain by Mengistu.

Mengistu remains unshaken in his conviction that Ethiopian
unity must be maintained, with substantial provincial autonomy
a danger to national survival.   He continues to ascribe
divisions within the nation as a whole to the perfidious policies
of the Emperor, encouraged by covert sympathizers. He looks
toward a nucleus of officers, dominant in the PMAC, COPWE
and now in the Central Committee of the Workers' Party, as
the key agents of national unity.   The mid-1970s rhetoric of
self-determination was far more the product of temporarily
coopted civilians (especially MEISON) than of governing
military personnel.   The Derg and its successor organs, all
under the thumb of Mengistu, have responded to cultural
diversity by insisting on the territorial integrity of the state.

Although many "nationalities" may exist, none enjoys the right of secession or self-determination broadly defined; at most, they can (when and if promised steps toward wider popular participation in politics are taken) work toward limited self-administration.

The armed forces have become the most important agents of national integration, imposed from above. Effective repression has fragmented the various regional rebellions into isolated pockets of resistance, although the total elimination of Eritrean, Oromo, Somali and Tigrean dissident groups appears highly unlikely. Contrasted with a decade ago, however, the Ethiopian government has met with extraordinary success. The power of the PMAC, under severe challenge in 1977-8, dramatically expanded with the massive influx of Soviet arms and of Soviet and Cuban military advisers, the conscription of an essentially new peasant army, and the exigencies of drought and famine that impacted most severely on rebellious areas. Whether coercion of this sort will achieve social integration in the long run obviously remains to be demonstrated.

Conclusion

Ethiopia is typical of African states in its cultural pluralism, yet distinctive in the historic continuity of its core. More than almost any other multi-ethnic sub-Saharan state, it maintained an indigenous political focus in the face of pressures for modernization. The 1974 revolution revealed to the world the sweeping contradictions and widespread exploitation of Haile Selassie's paternalistic system, an autocracy ill-suited to the emerging needs of Ethiopian society. Yet the subsequent groping for a new formula of legitimation has yet fully to succeed. Successive efforts - Ethiopia Tikdem, proclamation of socialism and Marxism-Leninism, establishment of COPWE, and formal creation of the Workers' Party a decade after the removal of the Emperor - have been acts in which hope, urgency and reaction have been mixed in about equal portions.

The central role of Mengistu has been played out through an increasingly limited segment of the military. Totally reequipped, dramatically expanded, and shorn of the leadership

of the imperial period, the Ethiopian armed forces have taken
on themselves the near-impossible goal of national unity. The
Derg's rise to power challenged and overturned the long-
standing political formulas; the installation of new norms
remains incomplete, however. Restive groups were encouraged
by the relaxation of old-style central controls and the debate
over "nationalities." As governing officers turned increasingly to
coercion, discontent rose. A vicious cycle seems in operation:
unity achieved through the armed forces has been gained
through suppression, which in turn has exacerbated regional
and ethnic fears. The quashing of one threat to the center,
with the forcible pacification of the Haud and Ogaden, cannot
mask the continuing resistance in Eritrea and the growing
discontent in Tigray. Quintessential Ethiopian characteristics -
and, it is tempting to write, Amhara characteristics - of
reducing peripheral groups to subordinate status mark the
current government. The methodology differs dramatically from
that of Haile Selassie; the objectives differ slightly. The national
interest of Ethiopia, as interpreted by armed forces, requires
national unity - and that unity (despite the rhetoric of the
Workers' Party) has at its core the primacy of the Amhara,
joined by coopted members of other groups (particularly the
Tigreans and Oromo). The undoubted accomplishments of the
revolution cannot conceal this essential continuity in Ethiopian
politics.

Notes

1. In 1980, there were an estimated 3.9 million Ethiopian
males aged 15 to 49 considered physically fit for military
service, with approximately 331,000 reaching draft age annually.
As Nelson and Kaplan note, Ethiopia has "far more manpower
than it could support logistically or train effectively" (Nelson
and Kaplan 1981: 256-7).

2. The formal expulsion of US Ambassador Frederic Chapin
ratified the break in American-Ethiopian relations inherent in
the revolution itself and its leftward course. However, I concur
with Halliday and Molyneux, who write, "Ethiopia was not

'lost' [to the United States]: rather, it was thrown away by the Carter Administration, which placed the continuance of relations with the Derg below such priorities as the courting of Somalia and the human rights policy. The 'loss' of Ethiopia had nevertheless begun well before 1977... Ultimately the USA 'lost' Ethiopia because the [US-Ethiopia] alliance depended upon the continuity of the absolutist regime: the failure of Haile Selassie's regime to resolve the contradictions within Ethiopian society... was the cause of both the 1974 revolution and the subsequent rupture in US-Ethiopian relations." (Halliday & Molyneux 1981: 225)

3. The precise reasons remain unclear, although demographic expansion and the spread of Islam through the jihad of Ahmed Gran seem central. For details, see Trimingham 1962.

4. One noteworthy early failure must be noted. In May 1976, a hastily-levied, essentially untrained peasant army of 10,000 to 40,000 members, largely Amhara and Tigrean in composition, was sent to the Eritrean border; two brief engagements led to "total disaster for the central government." (Ottaway and Ottaway 1978: 160) This debacle led to Mengistu's decision to raise a large, essentially new army from among peasants who had profited from the land reform, meaning far greater participation from southern Ethiopia. As a result, perhaps 60 percent of the army, and as much as 80 percent of the militia raised in 1977 to quell Eritrean separatism, came from the Oromo. (Halliday & Molyneux 1981: 196)

5. Throughout the continent, self-determination within former colonial boundaries has been the norm – witness the Organization of African Unity's stand on the former Spanish Sahara. Should Eritrea be considered an entity separate from Ethiopia? The African answer is historically clear: when the OAU was founded in Addis Ababa in 1963, the Ethiopian government sharply rejected Somali claims to the Haud and Ogaden areas. The second summit meeting of the OAU, held in Cairo in 1964, passed a strong resolution upholding the boundaries inherited at independence. The Ethiopian government is extremely sensitive about external discussion about ethnic cleavages and self-determination.

6. In the 1952-74 period, Ethiopia received $$118 million, compared to $34.8 million given to the 13 other recipient tropical African states. (Lefever 1970: 157)

References

Abate, Yohannis. 1984. "Civil-Military Relations in Ethiopia." Armed Forces and Society, 10: 380-400.

Abate, Yohannis. 1983. "The Legacy of Imperial Rule: Military Intervention and the Struggle for Leadership in Ethiopia, 1974-1978." Middle Eastern Studies: 28-42.

Abrahamsson, Bengt. 1972. Military Professionalization and Political Power. Beverly Hills, CA: Sage.

Africa South of the Sahara 1986. London: Europa Publications.

Baxter, P.T.W. 1978. "Ethiopia's Unacknowledged Problem: The Oromo." African Affairs 77, 308: 283-96.

Baynham, Simon, and Richard Snailham. 1983. "Ethiopia," in John Keegan, ed., World Armies. Detroit, MI: Gale, second edition.

Ehrlich, Haggai. 1983. The Struggle Over Eritrea 1962-78: War and Revolution in the Horn of Africa. Stanford, CA: Hoover Institution.

Gilkes, Patrick. 1974. The Dying Lion: Feudalism and Modernization in Ethiopia. London. Julian Friedmann.

___ 1983. "Centralism and the Ethiopian PMAC." In I. M. Lewis, ed., Self-determination in the Horn of Africa. London: Ithaca Press.

Greenfield, Richard. 1965. Ethiopia: A New Political History. New York: Praeger.

Halliday, Fred and Maxine Molyneux. 1981. The Ethiopian Revolution. London: Verso.

Harbeson, John W. 1979. "Socialist Politics in Revolutionary Ethiopia." In Carl G. Rosberg and Thomas M. Callaghy, eds., Socialism in Sub-Saharan Africa: A New Assessment. Berkeley, CA: Institute of International Studies.

Hess, Robert L. 1970. Ethiopia: The Modernization of Autocracy. Ithaca, NY: Cornell University Press.

Hoben, Allan. 1973. Land Tenure Among the Amhara of Ethiopia. Chicago: University of Chicago Press.

Huntington, Samuel P. 1957. The Soldier and the State: The Theory and Politics of Civil-Military Relations. Cambridge, MA: Harvard University Press.

Kirk-Greene, A.H.M. 1981. 'Stay by Your Radios': Documentation For a Study of Military Government in Tropical Africa. Leiden: Afrika-Studiecentrum.

Lefever, Ernest. 1970. Spear and Scepter: Army, Police, and Politics in Tropical Africa. Washington: Brookings Institution.

Lefort, Rene. 1983. Ethiopia: An Heretical Revolution?, trans. by A. M. Berrett. London: Zed Press.

Legum, Colin, ed., Africa Contemporary Record. New York: Africana Publishing Corporation, annual.

Legum, Colin and James Firebrace. 1983. Eritrea and Tigray. London: Minority Rights Group.

Levine, Donald N. 1974. Greater Ethiopia: The Evolution of a Multiethnic Society. Chicago: University of Chicago Press.

___ 1968. "The Military in Ethiopian Politics: Capabilities and Constraints." In Henry Bienen, ed., The Military Intervenes: Case Studies in Political Development. New York: Russell Sage Foundation.

___ 1965. Wax and Gold: Tradition and Innovation in Ethiopian Culture. Chicago: University of Chicago Press.

Marcus, Harold G. 1983. Ethiopia, Great Britain, and the United States, 1941-1974. Berkeley and Los Angeles; CA: University of California Press.

___ 1975. The Life and Times of Menelik II: Ethiopia, 1844-1913. Oxford: Clarendon Press.

Markakis, John. 1974. Ethiopia: Anatomy of a Traditional Policy. Oxford: Clarendon Press.

Nelson, Harold D. and Irving Kaplan. 1981. Ethiopia: A Country Study. Washington: U. S. Government Printing Office, third edition.

Ottaway, Marina and David. 1978. Ethiopia: Empire in Revolution. New York: Africana.

___ 1982. Soviet and American Influence in the Horn of Africa. New York: Praeger.

Perham, Margery. 1969. The Government of Ethiopia. London: Faber, second edition.

Scholler, Heinrich and Paul Brietzke. 1976. Ethiopia: Revolution, Law and Politics. Munich: Weltform Verlag.

Schwab, F. 1972. Decision-Making in Ethiopia. Rutherford, NJ: Fairleigh Dickinson University Press.

___ 1985. Ethiopia: Politics, Economics and Society. Boulder, CO: Lynne Reinner.

Selassie, Habte Bereket. 1980. <u>Conflict and Intervention in the Horn of Africa</u>. New York: Monthly Review Press.

Sherman, Richard. 1980. <u>Eritrea: The Unfinished Revolution</u>. New York: Praeger.

Trimingham, J. Spencer. 1962. <u>Islam in Ethiopia</u>. Oxford: Oxford University Press.

U. S. Arms Control and Disarmament Agency. 1988. <u>World Military Expenditures 1987</u>. Washington: U. S. Government Printing Office.

Welch, Claude E., Jr. 1977. "Warrior, Rebel, Guerrilla and Putschist: Four Aspects of Political Violence." In Ali A. Mazrui, ed., <u>The Warrior Tradition in Africa</u>. Leiden: Brill. London: Faber, second edition.

8

STATE-CONSOLIDATION AND SOCIAL INTEGRATION IN NIGERIA: THE MILITARY'S SEARCH FOR THE ELUSIVE

Stephen Wright

The military has proved to be very important in the sociopolitical life of Nigeria, having held the reins of government for all but four years since 1966. Such control has offered the opportunity to be involved in continuous efforts at social integration not only within the armed forces but also in the society as a whole. To some extent, the collapse and military overthrow of the civilian regime of President Shehu Shagari (1979-83) was symbolic of the failure to overcome the sociocultural cleavages which abound in the country, a task which the military set itself between 1966 and 1979. Continuing instability within the military itself since 1984 highlights the internal strains which serve to undermine the solidity and unity of the armed forces.

There have been many excellent scholarly contributions published on the role of the military and the reasons for its intervention in Nigerian political life,[1] but this chapter focuses upon attempts by the military to foster nation-building or rather state-consolidation - and social integration, specifically within its own ranks but also within the wider society. The period of study is from 1960, the year of independence, to

179

1990; during this 30 year period there have been six coups d'etat (two in 1966; 1975; 1976; 1983; and 1985) and several unsuccessful attempts. The discussion here focuses both upon the relative success or failure of the integrative process, and also upon the reasons why such a process is necessary to promote stability within Nigeria, where questionable loyalty to the state and the presence of extreme sociocultural cleavages exist. In addition to the general issues raised within the frame of reference for this volume, the analysis presented here will be extended to consider whether two decades of military leadership have promoted a greater level of social integration, and implicitly successful government, than civilians could have offered, and whether the proposed transfer of power back to civilians in October 1992 will have a significant impact on sociopolitical factors or whether it will merely constitute a breathing space before a further military incursion into the political arena takes place.

Before discussing these issues, it is important to consider the major sociocultural cleavages which dominate Nigerian life and permeate the country's armed forces, so making the task of social integration an immensely difficult one.

Sociocultural Cleavages in the Nigerian State

Nigeria provides an excellent example of the artificiality of a modern African state. Its sociocultural cleavages are so great that one could speculate that integration of its diverse groups is an almost impossible task. Within the context of this chapter, integration is defined loosely as a process to minimize or eradicate divisive social cleavages between identifiable subgroups in an attempt to foster some form of national identity, spirit and unity.

There are over 250 ethnic groups and languages in the country, although English serves as a common and, to a certain extent, unifying language. The British colonial authorities administered Nigeria in three specific regions: the North comprised approximately half of the population, while the West and East[2] together accounted for the other half.

Because political constituencies and revenue allocation within the postindependence federation are shaped by demographic considerations, censuses have traditionally been extremely contentious, and the disputed results have served to increase interethnic hostility.[3]   The regional configuration of the colonial regime was adopted in the First Republic (1960-66) as the three governments were dominated by the ethnic group predominant in the region: the Hausa-Fulani in the North, the Yoruba in the West and the Ibo in the East. In 1964, a fourth region, the Mid-West, was created. The cleavage between the regions was reinforced by religious differences, with a majority of northerners being Islamic and southerners largely Christian, though a number of pagans and Yoruba Muslim made these divisions less stark.

Sociocultural cleavages within the polity proved extremely destabilizing in the early 1960s and, combined with the strains of rapid "modernization" [4] and the breakdown of the traditional social order, culminated in the collapse of government in the West after 1964 and the overthrow of the northern-dominated federal government by the military in January 1966 (Post and Vickers, 1973). During the next 12 months, ethnic conflict was open and vicious with some 50,000 Ibo civilians murdered in the North and over one million of them fleeing back to the relative safety of the East. In May 1967, the Eastern region declared itself a secessionist state, Biafra, and the majority of the Ibo population opted out of the Nigerian federation. The subsequent civil war did not end until the Biafrans surrendered to federal forces in January 1970. Although the then head of state, General Yakubu Gowon, was relatively successful in reintegrating Biafrans, the memories and scars of that war continue to linger in Nigerian minds.

Ethnoreligious cleavages have become less intense since the early 1970s as a result of a number of factors. Successive military governments have broken up the four regions into 12 states in 1967, then into 19 states in 1976, and 21 states in 1987. This action served to lessen tensions, and has provided greater stability, although it has not prevented claims for the creation of more states in the federation. An important repercussion of these actions was to allow the minority ethnic groups of the Middle Belt states in the geographical center of the country to emerge from the shadow of the predominant Hausa-Fulani and so gain increased political influence to correspond with their relative numerical strength within the armed forces. The division and demise of the ethnically-oriented regions (in particular the North) also helped to increase the federal government's control over the states, a trend which was reinforced in the late-1970s by the burgeoning oil revenues which were concentrated in federal hands and made the states financially dependent upon the government in Lagos.

The military government of Lt-General Olusegun Obasanjo (1976-79) made a conscious effort to diminish ethnic rivalry before the return to civilian rule in October 1979 by introducing strict criteria emphasizing the national rather than ethnic composition of the political parties which were allowed to contest the 1979 elections. These restrictions proved to be marginally successful: the winning party, the National Party of Nigeria (NPN), was a little more "national" in outlook than its predecessor of the 1960s, the Northern People's Congress (NPC), but it was still dominated by Hausa-Fulani elements. The other four legalized parties similarly reflected residual ethnic sentiments within superficially national frameworks. All the parties (except perhaps the People's Redemption Party) were bourgeois and conservative in outlook, reflecting the military's own political perspectives, and the military barred more than 20 parties, including all the socialist groups, from contesting the elections (Falola and Ihonubere, 1985). The failure to eliminate ethnic considerations contributed to the eventual removal of the civilian government in December 1983 (Laitin and Harker, 1981; Diamond, 1982), and ethnicity remains a considerable issue to be dealt with in the demilitarization exercise before 1992.

Religious tensions within Nigeria have been fairly well controlled, or perhaps avoided, as a result of a considerable level of tolerance of religious differences. There have however, been a number of potentially explosive issues over the last decade which could continue to prove troublesome.[5] The debate surrounding the formulation of the new constitution in 1977-78 foundered for a period over the status of the Islamic shari'a courts within the federation, an issue which proved to be the most controversial of the whole constitutional exercise (Laitin, 1982). This shari'a issue again proved to be highly contentious in 1988 in the constitutent assembly's debates concerning a new constitution for 1992, and the military government was forced to ban discussion of this issue. A second controversy has been Christians' concern at their inability to establish churches in the Islamic north. Although not officially restricted, Christians have found it increasingly difficult to buy land and obtain planning permission for buildings, and consequently complain bitterly of petty discrimination. A third factor focuses on Nigeria's accession to the Organization of Islamic Conference (OIC) in January 1986. Christians vehemently protested that this violated the secular character of the country, and despite the military government's official concern, membership was maintained. This issue still simmers, and remains a focus of controversy.

Social Pluralism in the Military

The intense sociocultural divisions within Nigeria have had a marked impact upon the country's armed forces. This was particularly the case in the early years after independence, but these cleavages still continue to exert influences on the military today. Awareness of the significance of such sociocultural factors has led to a number of specific policies being intro- duced, many of which are discussed below, but has also led to the conscious policy of eliminating or obscuring data or information which focus upon these rivalries within the army, so making it difficult to formulate an accurate profile of the armed forces.

The army remained very small in total numbers in the early years after 1960, as the colonial Royal West African Frontier Force (RWAFF) of 7,500 men at independence grew to some 12,000 troops by 1966 (Miners, 1971). The serious ethnocentric problems of political life were reflected in the composition of, and divisions within, the armed forces. At independence, over three-quarters of the officers were expatriate, while the small number of African officers was made up almost entirely of Ibo personnel. The highest-ranking indigenous officer was a major. In contrast to the ethnic composition of the officer corps, the bulk of the troops emanated from the Northern region in general, and from the smaller ethnic groups of the Middle Belt area in particular.

The complete Africanization of the officer corps took place in the short period between 1961 and 1965, a process which generated intense rivalry among the ethnic groups. The federal government, dominated by northern interests, introduced in 1961 a quota system attempting to recruit 50 per cent of all officers from the North, and 25 per cent each from the West and East. People in these latter two regions pressed unsuccessfully for open recruitment without quotas on the basis of educational ability, knowing full well that if this were adopted they would benefit greatly over the educationally underprivileged North.

The military's overall impact on the political process was minimal at this time. As a very small force with obsolete equipment and little prestige because of its absence from the nationalist struggle (except perhaps to quell protests), there was little popular identification with the army. The rapid Africanization process and promotion of indigenous officers contributed to a lack of cohesion and esprit de corps, or what has been labelled "professional disorientation" (Gutteridge, 1975). Few people were concerned by the internal instability within the army as nobody presumed that the military had the willingness or ability to intervene in political life. [6] Even an authority such as William Gutteridge (1965: 107-8) commented that, "In terms of influence on the affairs of state...the army is less likely to play a decisive role than (it is) in other new states."

Even though the coup d'etat in January 1966 by army majors came as something of a surprise, its impact on the country was devastating. A number of pertinent factors needs to be discussed to explain the coup. Despite the quota system of recruitment and many politically-inspired promotions (Odetola, 1978), a significant majority of officers was of Ibo origin at the end of the Africanization process in 1965. This placed the military leadership out of balance with the political leadership in the federal government. The Ibo officers feared that after a period of rapid promotion during the Africanization process, there would be greatly restricted mobility in subsequent years. On a political level, Ibo officers were known to be resentful of the domination of the federation by northern politicians and by the inability of non-northerners to make their voices heard.

Although the coup instigators of January 1966, led by Major Chukwuma Nzeogwu, claimed nationalistic sentiments in their move against a feudal, ethnically-motivated government, this blow to northern supremacy triggered off a rapid destabilization of the country. Northerners refused to see the army's intervention in anything but ethnic terms, even though northern troops had followed the orders of their officers in the coup. The assassination of the two preeminent northern leaders, Sir Alhaji Abubakar Tafawa Balewa, the federal premier, and Sir Alhaji Ahmadu Bello, the northern premier, combined with the almost exclusive Ibo orientation of the new military government, were serious factors. Equally, the abolition of the federal structure and the imposition of a unitary governmental system in May 1966 was viewed not as an attempt at state-consolidation or nation-building, but as a means to undermine and suppress the sociocultural identity of the North. The serious rift widened within the officer corps and culminated in a counter-coup by northern officers in July 1966. Ethnic hostility inside and outside the army reached fever pitch, and these strains effectively split the federal army into two, with Ibo officers and troops forming their own Biafran army to fight the rump federal forces (Miners, 1971).

At the end of the civil war in 1970, over half of the Biafran officers were successfully reintegrated into the federal army, a dramatic if somewhat isolated example of nationalism overriding ethnicity. Such a gesture did not extend to the rank and file Biafran soldiers who were forced to disband. This reconciliation also did little to help Ibo officers get back their position of influence within the armed forces. Resentment and bitterness remain apparent even in the late 1980s, because of the absence of Ibo officers from military governments: not a single representative on Major General Muhammadu Buhari's Supreme Military Council (SMC) in 1984-85, and only one officer in Major-General Ibrahim Babangida's initial Armed Forces Ruling Council (AFRC) from August 1985.

A number of other serious repercussions of the civil war on the military can be identified. The first concerned the size of the army, which had grown dramatically during the war from 12,000 in 1966 to 270,000 in 1970. The civilian leadership prior to 1966 had kept the army small to minimize the chances of a coup as well as to limit the possibility of internal struggles between powerful rival ethnic subgroups. Fears of such potential actions within the enlarged armed forces were also held by military leaders in the 1970s. Another problem concerned the cost of the army, whose bloated size led to it taking one-third of the total federal budget in 1974, 90 per cent of which went towards the payment of salaries (Legum, 1975). Although there was a consensus that the army was too large, there was little willingness to attempt demobilization. The belief was widely held that such a process would cause severe societal strains following the emergence of a large number of semi-educated, yet militarily capable, people destabilizing the delicate political framework of the 1970s (Gutteridge, 1975). General Yakubu Gowon (1966-75) shied away from this task, but his successors, General Murtala Muhammed (1975-76) and General Olusegun Obasanjo (1976-79), grasped the nettle and trimmed 100,000 troops between 1975 and 1979, the year in which the transfer of power to civilians took place. [7] By 1989, Nigeria's total armed forces had been cut further to approximately 90,000.

Following the establishment of the Second Republic in October 1979, the civilian leadership under President Shehu Shagari (NPN) set out to control the sociocultural cleavages within the armed forces, as well as to minimize the chances of a coup. Shagari tackled the problem politically. He demoted or removed a number of officers, largely Yoruba, whom he perceived to pose a potential threat to his government, and at the same time hastened the promotion of officers of northern origin. The president also increased wages and overall military spending in an effort to pamper the military into willing subservience (Adekanye, 1985). The strategy did not work, and the government was overthrown in December 1983 mainly as a result of political and economic incompetence and gross corruption (Diamond, 1982; Falola and Ihonvbere, 1985). The coup was masterminded by northern officers, a fact which on the surface appears to be out of sequence with political events, even though there is significant factionalism within the northern ranks. This can be explained by the fact that there were widespread rumors at the end of 1983 of an impending coup by junior officers drawn largely from southern Nigeria. If such a coup had taken place this would have upset the delicate sociocultural balance within the army, as had the 1966 coups, and would also have undermined the command structure within the armed forces. It has also been speculated that the leaders of that embryonic coup could have favored radical policies, following the role models of Jerry Rawlings in Ghana and Samuel Doe in Liberia, and so could have challenged the economic status quo favored by senior military leaders.

It is a significant fact that sociocultural cleavages within the 90,000 man armed forces (which comprise 4,000 in the navy and 9,000 in the airforce) are not as serious in the early 1990s as in the first decade after independence. This is a product of the positive factors of social integration as well as certain negative issues, such as the fear of a repetition of the chaotic and bloody conditions of the mid-1960s. A "palace coup" of August 1985 removed Muhammadu Buhari as head of state (he was finally released from house arrest early in 1989) and ushered in Ibrahim Babangida, an indigene of Niger State in the Middle Belt. Babangida has consciously worked to strengthen Middle Belt influence and representation within the Armed Forces Ruling Council (AFRC), reflecting the fact that some two-thirds of the armed forces come from minority ethnic groups in this region, and has consequently weakened northern influence over the military and the country as a whole. This attempt at social engineering is fraught with hazards, especially in its direct challenge to northern vested interests. Furthermore, the Middle Belt itself is by no means an homogenous unit, but is merely a label used to identify smaller ethnic groups who stand in the shadow of the larger Hausa-Fulani, Yoruba and Ibo groups.

Two important events which took place soon after the military changeover highlighted the vulnerability of Babangida's government in the face of these sociocultural pressures. The arrest of more than a dozen military conspirators in December 1985 (and their subsequent execution) rocked the new government, especially as the alleged leader of this group, General Mamman Vatsa, was a trusted ally from the Middle Belt state of Benue. [8]   The second issue has already been touched on, namely the accession to the Organization of Islamic Conference.   This decision was highly controversial within the AFRC as only the Muslim members were proxy to the decision, and Christian officers did not know of it until informed by the newspapers. [9]   Such blatant religious overtones within the highest ranks of the military served to increase cleavages between groups, and may well become a factor in future actions by disgruntled members of the armed forces.

Socialization Through the Military Experience

It is evident that sociocultural cleavages have proved to be significant obstacles to state-consolidation throughout the period since independence, and that these have had a major impact upon the military's stability and orientation.   It is worthwhile now to shift the focus of discussion to consider whether the military experience helps to promote social integration. Initially, the focus is upon whether military experience itself changes the social and national values of those involved, while the second section considers the impact of specific policies instituted by the military leadership to foster social integration.

The Nigerian military, like militaries in many Third World countries, serves as a useful socializing mechanism and has achieved a considerable degree of unity overriding the enormous sociocultural divisions present in the country.   The serious disunity of the civil war era was overcome to a considerable extent by more balanced policies in the 1970s, an occurrence also helped by the shared experience of the bitterness of that war.   Tensions present within the military in recent years have been generated perhaps more from frustration with the poor performance of its leaders, as evidenced by the explanations given for the overthrow of Gowon in 1975 and Buhari in 1985, than by more traditional sociocultural cleavages.

Several issues need to be raised when attempting to weigh the social impact of military service. Three fundamental points can be made immediately. The first is that successful integrative efforts within the armed forces do not automatically foster beneficial spin-off effects on civilians for reasons outlined below. The second factor concerns the exclusion of women from the Nigerian military: with 50 per cent of the population excluded from the military process on grounds of sex, the overall impact on society of the military experience is diluted. The third offers a radical perspective that military leaders, as members of a bourgeois elite, gain directly and personally by controlling government, and that some may wish to limit the success of social integration efforts. The resultant effect is that ethnic groups are manipulated to remain pitted against one another in a form of zero-sum game, and this partly-contrived hostility hampers the development of a national working class movement which would threaten the status quo (Falola and Ihonvbere, 1985).

Compulsory national service has always been considered too costly and potentially dangerous in a sociopolitical sense, and so the armed forces are a volunteer force. Salaries are not high for any of the service levels, and consequently the majority of the recruits (though not necessarily officers) tend to be semi-educated. It is impossible to gauge with accuracy the size of Nigeria's total population, as no truly accurate census has been possible since independence because of political manipulation. However, most guestimates cluster around the continuum of 100 to 120 million. On this basis, the military accounts for approximately 0.001 per cent of population, or the ratio of 1:1,000, a relatively insignificant figure when judging the impact of the integrative process within the military and its impact on society as a whole.

The potential influence of military service is further minimized when one considers that it is essentially regarded as a long-term career commitment, and so the demobilization process does not occur at a rapid pace or with significant numbers such as it would with a short-term conscript force. Added to this is the fact that there is generally a low level of interaction between military personnel and local civilian communities, with troops rarely being involved in development schemes. In the period of reconstruction following the civil war, for example, such projects appeared ideal for the military to undertake, but troops were seldom used (Odetola, 1982). Several factors help to explain this absence of linkages. On the one level, it is widely believed that troops do not possess the skills necessary to be useful in such development projects and, where used, their assistance cannot be depended upon. On the other hand, officers have consciously worked to keep troops detached from civilian projects for fear that the soldiers will become contaminated by socialist/workers' ideals.

Another factor to consider concerns the behavior of the military and the example that it sets for other sections of society. This reputation is by no means favorable at officer or rank level. Numerous skirmishes have been recorded up to the present day in which uncontrollable troops have attacked civilians, wrecked vehicles, and acted in a lawless manner, and these events obviously serve to lessen respect for the armed forces. Even if there was impeccable discipline, this would not necessarily improve the integration or command structure of the army; as Lee (1969: 145) pointed out, "acceptance of discipline does not automatically imply a recognition of authority."

Corruption, defined here simply as the misapplication of public funds for private purposes, is a national pastime of Nigeria and has afflicted military and civilian regimes alike. Probably the high point of acknowledged corruption in military government came near the end of General Gowon's period of office, between 1973 and 1975, when embezzlement of funds was overt, and when a series of corrupt deals brought so much cement to Nigeria that in 1974 a total of 450 ships loaded with cement lay off Lagos harbor waiting to dock. [10] The "great purge" of General Murtala Muhammed lasted less than nine months (1975-76), but removed 10,000 civil servants as well as 150 military officers. This redeeming zeal by a man himself accused of corruption was widely approved by the majority of Nigerians, but not by a group of disgruntled officers who assassinated Murtala in February 1976, allegedly with the knowledge of the British Government and the connivance of the U.S. Central Intelligence Agency (Falola and Ihonvbere, 1985). [11]

Military governments during the 1970s only marginally

improved Nigeria's position within the international economic system, but they did not provide any measurable benefits for the majority of Nigerians through economic development programs nor through the indigenization decrees of 1972 and 1978. Most observers concur that the military squandered the opportunities presented by the very considerable oil revenues, which rose rapidly during the decade and which peaked at $25 billion in 1980 (Dudley, 1982; Ojo, 1984).

The failure to instill the concept of probity in the civilian elite was evident when the Shagari government (1979-83) was estimated to have embezzled in the region of $5-7 billion. The subsequent military regime led by Muhammadu Buhari did little to improve the military's image, but Babangida's government since 1985 has attempted to lessen overt corruption. However, Babangida has had to work against chronic economic conditions, with Nigeria's foreign debt up to $29 billion in 1989, and oil earnings down to below $6 billion. But honesty in government is generally a precious commodity. As Chief Obafemi Awolowo, one of Nigeria's foremost political figures prior to his death in 1987, lucidly pointed out, "since independence our governments have been a matter of a few holding the cow for the strongest and most cunning to milk" (Dudley, 1982: 225).

Successive military leaderships, then, have often failed to give an effective lead to society and have been themselves guilty of many misdemeanors. Consequently, the military offers limited "symbolic" value to Nigerians as a model to follow in the state-consolidation exercise, as many of the corrupt and ethnocentric practices discussed earlier serve to undermine further the prestige and status of the army. The difficult task of governing the country has brought little credit to the military, and has in fact often dragged it deeper into the quagmire of divisive sociocultural politics. The military did take a symbolic lead in 1985 by imposing salary cuts of 2-15 per cent on itself to lead the country's austerity program, but this gesture lost its validity when similar cuts were imposed on the whole public sector. The armed forces have also not gained prestige from their external military actions, having muddled through after border skirmishes with Benin and Cameroon in the early 1980s, and failing in their direct intervention in the Chad conflict in 1982. Given the fact that Nigeria's military is the second largest in black Africa, and that the country accounts for approximately a quarter of the continent's total population, these failures helped to diminish further Nigerians' respect for their military.

To conclude this section, then, it can be said that military service does provide some form of socializing mechanism, but its level of success and the impact on wider society falls far short of what could be anticipated.

Integration Within the Military

In order to create greater possibilities for successful social integration within the armed forces, various policies have been purposefully introduced by both military and civilian governments.

Most of these efforts center upon attempts to minimize ethno-oriented suspicions and so help to forge stability and unity. To some extent these attempts have been successful in that the tensions prevalent in the 1960s have been subdued. Nevertheless, there remains considerable internal tension which requires astute leadership to control. Some of these potential problems present in the early 1990s can now be considered in discussing politico-military relations and the attempt to demilitarize the government apparatus.

As in other chapters of this volume, several issues are addressed. The first concerns recruitment. There has been a number of deliberate initiatives taken to reflect national character and composition in the armed forces, but these have never gained full support. The explanation for this failure rests largely upon ethnic sentiment. A "national" policy was not always favored, and disagreement has repeatedly occurred over the word's definition. In the early 1960s, as mentioned above, the more modernized and educated southerners pushed for open recruitment to the armed forces on the grounds of ability, knowing that this would give them a lion's share of the officers' positions. The North, which had attempted to stay aloof from the onslaught of "modernization" and Christianity and which had been encouraged to do so by the colonial policy of "indirect rule", believed that such a recruitment policy would give its people little chance to move into positions of responsibility within the army. As the North's numerical strength (based on dubious census results and including women who were actually denied the vote until 1979) translated via the ballot box into political influence, so its interpretation of "national" was introduced in the "quota" system of 1961. Under this scheme, national was interpreted to mean a reflection of relative ethnic size and, in effect, to allow the opportunity for the North to catch up to the South in terms of officer recruitment.

As we have already seen, such intense ethnic rivalry broke the armed forces into two camps in 1966. Following the civil war, a period of healing and reconciliation leant weight to integrationist moves within the army. Northern leaders had come to positions of leadership during the war, and in the 1970s a rapprochement developed between Hausa-Fulani and Yoruba elites. Following the assassination of Murtala Muhammed, a Hausa-Fulani, in February 1976, the peaceful transition of power to the second in command, Lt-General Olusegun Obasanjo, a Yoruba, was a favorable indication of the relative progress towards the goal of social integration within the military.

The reemergence of a northern-based civilian government in 1979 brought questions concerning ethnicity and quotas within the armed forces to the surface again. Possessing the memory of having been pushed aside by an Ibo-dominated military elite in 1966, northern political leaders attempted to protect their careers by implementing a number of policy decisions. Many of these were adopted subsequently by the post-1983 military governments and so need to be discussed in some detail. One novel approach was the clause inserted in the 1979 constitution, with the military's approval, which forbade any future military incursion into the political arena.[12] This gesture was based on the belief that the military did not possess any inherently superior national characteristics and, by its intervention into the political arena, actually retarded social and political development.

The Shagari administration officially based its recruitment

decisions on "federal character", a vague yet promising concept drafted into the 1979 constitution which meant that appointments should reflect the cultural diversity of the country. This criterion was to apply not only to military appointments but also to the federal civil service and all other national recruitment agencies. Such a policy reopened the can of worms concerning "quotas" or "open educational" appointments and promotions, and tended to undermine promotion by ability. The federal government displayed a bias towards the promotion of northern officers, although not in terms of strict numerical quotas, and attempted to win the officers' support for the regime. Because many of the senior officers were, and remain, from the Middle Belt states rather than the far North, tensions emerged over this policy. These minority groups are now relatively well represented in the Babangida administration, but tensions between the various groups remain.

President Babangida is fully aware of the problems of balancing ethnic rivalries within the armed forces between the dominant and minority groups. Northern interests appeared to be in the ascendancy under his predecessor, Muhammadu Buhari, who was linked to the influential northern business elite often referred to as the "Kaduna mafia", but the Middle Belt dominated regime of President Babangida has shifted the power center (temporarily) away from the North. Paradoxically, the dangers inherent in the mechanics of ethnic balancing appear to have a beneficial impact on the military leadership, which is forced to increase its efforts to foster social integration, even though the terms of reference may differ according to the perceptions of the various factions. What is important to note is that the ethnoreligious composition of the AFRC is monitored closely by the population and helps to set the tone for societal views on integration -- too heavy a concentration of any one ethnic group within the government can provoke serious repercussions.

The military introduced specific policies after the civil war to integrate its personnel when efforts were made to break down the regionally-based armies of the 1960s. Unit commanders were shifted around the country at least every two years, both to provide awareness of the social dynamics of other peoples and regions, and also with the practical intention of trying to prevent a close attachment with troops at any particular base which could facilitate the planning of a coup. To promote national integration, no units were allowed to be monoethnic, although many had high concentrations of one ethnic group in the region in which they were located. In the 1970s, a number of new military bases was established in towns which previously lacked a military presence. This helped to dilute the previous concentration of military power with bases in only six centers, which had been the pattern in the 1960s, but also helped to give a greater national orientation to the military. It also made a coup attempt more difficult to undertake because of the need to move on a greater number of locations simultaneously (Panter-Brick, 1978).

Education programs, especially for officers, were stepped up in the 1970s and numbers of personnel sent to the Command College, Jaji, near Kaduna, increased dramatically. Such opportunities were officially based on nonethnic criteria, and these programs have undeniably contributed to greater unity within the junior officer ranks. In the second half of the 1980s, restrictions were placed on the number of officers being sent overseas for training, a policy laced with the political rhetoric of independence and nationalism, but underwritten by the harsh economic realities of the lack of foreign exchange. [13] In an attempt to readjust to these new facts, the Nigerian Defense Academy was upgraded to degree-awarding status in September 1985 and the number of places at Nigerian military institutions was increased. It remains too early to conclude what the exact long-term implications of these moves are. The contention by authorities that officers will be more nationally oriented and patriotic because they have undertaken their training in Nigeria rather than abroad is an interesting, but perhaps spurious, assertion.

Other educational policies for Nigerian military personnel and families are limited. What programs exist stress national over ethnic considerations, but this philosophy is not unique by any means, and is a standard policy stressed elsewhere, such as in schools and universities. Despite these efforts, ethnocentric loyalties remain predominant within society. The military is not isolated from these influences, and is having problems suppressing them. The ranks remain loyal to their ethnocentric linkages through culture, religion, political manipulation and language (despite the use of a lingua franca, English) and integrationist educational policies have to date been only nominally successful. The officer corps is more attuned to a national orientation, but nevertheless retains some ethnic sympathies.

Successive governments have made little use of retired military personnel to help bolster national unity. Those most visible, the retired senior officers, normally gravitate to business, where they are assumed to continue with the corrupt use of contacts gained during their military service. A few officers, such as Benjamin Adekunle (the infamous "black scorpion" of the civil war) and Olusegun Obasanjo, have used their status to encourage and lead the national debate concerning changes to the constitutional framework before the return to civilian rule, although their inputs have not always been exactly "national" in outlook. Obasanjo, who increased his prestige by his presence on the Commonwealth's "Eminent Persons Group" to South Africa[14], commented in 1986 that what mattered were the policies, rather than the nature, of government, indicating that the continuation of military rule could be good for Nigeria. This comment was reminiscent of one made by Obasanjo a decade earlier, when he was head of state, that the new civilian political system after 1979 should be devoid of political parties, an indication that he was aware the parties would be ethnically based. Adekunle, among others, raised points favoring the break-up of the federation into a confederation, effectively questioning the viability of Nigeria as a single entity. This policy, if ever implemented, would have obvious political implications, but the military aspects would also be very significant. At the beginning of 1967, immediately prior to the civil war, a similar idea had been mooted for a confederation of the four regions, each with its own ethnically-identifiable army. Such a proposal was a reflection of the strife in the country at that time, but was overtaken by the outbreak of the civil war.

The recent discussions over confederalism were officially closed in mid-1987, when the military government decided that a federal structure would remain in place after 1992. However, the revival of such an idea, especially when led by a prominent retired officer, raises doubts not only about the level of social integration achieved within the armed forces to date, but also about the possible breakdown of this tenuous unity as important decisions on the constitutional and political structure of the Nigerian state are made in preparation for the transfer of power to civilians in 1992 (Suberu, 1988).

Military-Civilian Relations and Social Integration

Although attempts by the military leadership to inculcate national sentiments within the armed forces can be considered of limited success only, this has not deterred efforts to implant a quasi-military ethos within wider society and instill similar ethical and national values. Two schemes deserve mention: firstly the National Youth Service Corps (NYSC), introduced by General Yakubu Gowon in May 1973, and the War Against Indiscipline (WAI), started by Maj-General Muhammadu Buhari in March 1984, reconstituted into the National Orientation Movement (NOM) by Maj-General Ibrahim Babangida in July 1986, and then again revamped as the "mass mobilization for economic recovery, self-reliance and social justice" (MAMSER) in 1987. Both the NYSC and WAI/NOM/MAMSER derived their existence from the military's belief and assumption that society's problems could be best solved by providing ordinary Nigerian citizens with a dose of military discipline, rather than by concentrating on trying to make the behavior of the country's political and military elites less corrupt and more socially acceptable. [15]

The NYSC, like its contemporaries in Ghana, Zambia and Tanzania, aims to stress and encourage national principles in the youth of the country, and thereby help to limit sociocultural cleavages. All university graduates are drafted into the scheme, despite their occasional protest, and are required to undertake one year's service prior to employment on behalf of the nation in areas as diverse as teaching to road-building. A principal strategy of the program is to assign recruits to serve outside of their home state, and preferably geographic region, in order for them to be exposed to the differing cultures within the country. This policy rests upon the supposition that acquaintance precedes acceptance. Unlike in other African countries, however, the NYSC scheme has no overt military implication. The NYSC program has had mixed successes over the 15 years of its existence. Many students, possibly up to half, are able to swing appointments close to home. Many of those who went grudgingly to far away locations - including a number interviewed by this author - returned home after the year with somewhat negative memories of their "exile", raising the possibility that the scheme could be counterproductive in some cases. Even the leaders of the scheme occasionally proved to be lacking in character: the NYSC chief who served in the Shagari period was jailed for embezzlement by the Buhari/Babangida governments.

The War Against Indiscipline (WAI) is a wider-ranging program, calling for participation by the whole country. The underlying assumption of this scheme is that Nigerians lack moral fiber, and that a national program using quasi-military methods will lead to a miraculous moral rejuvenation and greater national awareness and values. An initial plank of this strategy was to coax Nigerians to wait in line, the "queue", while other goals included cleanliness in towns and homes, patriotism, and the abolition of corruption.

These policies were initially met with enthusiasm, but then petered out at the hands of a somewhat cynical and astute population overburdened with economic hardship. The lack of success of the WAI scheme led President Babangida to introduce the revamped NOM in mid-1986, and then MAMSER in 1987, but these also made little headway. The implications of these failures are significant. It is clear that the average Nigerian needs more than a public lecture to change her/his behavior and allegiances, and that a "national service" approach hardly holds the key to change. It is also evident that people have little sympathy or respect for their military leaders, who themselves have been exposed on many occasions as lacking the nationalistic outlook which they are trying to press on others, and whose policies are often considered to be responsible for the chaos and hardship of Nigerian life.

This apparent failure to build meaningful bridges to society can be traced to other factors also. It is safe to conclude that the population does not share the military's opinion that the latter is best equipped to rule the country, especially for long periods. [16] It is true that the breakdown of the civilian political systems in 1966 and 1983 enabled the military "saviors" to be popular for a short time, but the basic philosophy of government by civilians is widely held.

The military's relationship with specific civilian groups is mixed. While not enthusiastically welcomed by any Nigerian, military rule is supported by some industrialists and businesspersons, if only because of the basic political and economic stability, or rather control, which such rule has generated in the past. Labor groups, on the other hand, have a difficult time with military governments, which appear to be unable to accept complaints and challenges to their authority. The Babangida government is continually at odds with the Nigerian Labor Congress (NLC) for many reasons: because of the absence of dialogue; the imposition of salary cuts; the failure to combat retrenchment with new jobs in the beleaguered economy; the problems inflicted on the working poor and underemployed by devaluation, the ending of food/oil subsidies and the privatization of state-owned corporations; the refusal in 1987 by the military to pursue a socialist national strategy; the dismissal of the complete NLC Executive by the government in 1988; and the frequent detention without trial from 1986 to 1989 of many of the NLC's leaders, despite the apparent human rights predeliction of the AFRC. The NLC and its radical supporters point to these facts to highlight the class bias of the military leadership and its unwillingness to pursue true structural economic reform, as opposed to the tepid structural adjustment programs supported by the World Bank and the International Monetary Fund.

Links with other significant social groups are also not too favorable. The Christian churches, acting through the Christian Association of Nigeria (CAN), have been opposed to the military's harsh policies to combat crime, including public executions for many offenses. [17] By African standards, the press has fared well under military governments, but there have been examples of breakdowns in relations. This happened in

1974 with widespread media criticism of corruption and inactivity under Gowon, while relations deteriorated seriously in 1984 when Buhari introduced Decree No. 4, which severely curtailed press freedom.[18] This decree was lifted by Babangida in August 1985, but the press remains cautious over its degree of freedom, and current discussion concerning the return to civilian rule is carefully monitored by the government.

Relations between the military and the universities have never been good, but have also deteriorated since 1984. The Nigerian Security Organization (NSO), established by General Murtala Muhammed in 1975, ran amok on the campuses in 1984/85, arresting a number[19] of faculty and students and, unwittingly, often its own members. Babangida stepped in to deal with the excesses of the NSO and reorganized its structure and activities, but there is little to suggest that Babangida's token moves to control the security services have increased support on campuses. To many students, the sight of the army or armed police on campus is a common one, and brutality against students, such as the killing of over 20 people on the Ahmadu Bello campus (Zaria) in May 1986, hardly helps to win converts to the military's cause. Following major riots in 1988 to protest fuel price increases, both the student and academic staff unions were banned, again serving to raise tensions.

Civil-Military Relations: Towards 1992

The final area to consider is the relationship between the military leadership and civilian political elite in the light of the debate concerning the transition to civilian rule by October 1992. There is little evidence to support the hypothesis that military governments (1966-79 and 1984 to date) have achieved more than civilian ones could have in terms of providing for economic growth or development, however defined, or for improving the quality of life of the average Nigerian. The "nation" of Nigeria was only marginally more stable in 1980 than it was in 1960, even after close to 13 years of military rule. The military also could be accused of being "parasitic", especially during Gowon's term of office in the early 1970s in maintaining very high costs and being extravagantly corrupt, causing an excessive drain on the country's finances.

The failure of the Second Republic to survive longer than four years illuminates the problems associated with demilitarization of the governmental machinery,[20] and raises constitutional and structural questions which the military leadership is currently attempting to solve. A major decision was made by President Babangida in July 1987 to impose a five-year ban on political activities on all senior politicians and party representatives of the Shagari era. Although many civilians worked hard to secure exemption from the ban early in 1989 prior to the legalization of political activity in May, the majority will not be allowed to contest the upcoming state or national elections.

This decision by the AFRC to prevent a return by these prominent politicians is indicative of the military's optimistic assumption that certain sociopolitical problems can be eradicated by decree. A second flaw in the military's plan is to believe that a new civilian elite will automatically emerge who espouses national, rather than sectional, sentiments. It is true that such an event is possible, but Nigerian political history tells us that it is unlikely. New leaders will probably be forced to seek legitimation from old-guard political figures and will be faced by similar pressures to rely for support on traditional, ethnic roots as they run for office, and so ethnic divisions could again be exploited rather than healed. The military allowed only two political parties to be registered to contest the elections in 1990/91, hoping to minimize ethnic conflict and force politicians to seek national mandates and coalitions. Internal conflict within the two parties in 1990 showed that national consensus would prove difficult to achieve. There is also a possibility that the new generation of leaders will be as corrupt as previous ones, and the "get rich quick" mentality (before the military intervenes) could dominate their thinking. There is little evidence to suggest that the politicians of the 1990s are significantly more nationalistic than those of the 1980s. Such a conclusion reflects poorly on the military's attempts to instill national values upon society over the last 20 years.

If the impact of the military's efforts at socialization and integration is considered marginal on wider society, and on politicians in particular, what conclusions can we draw about the military itself in terms of its unity and nationally-oriented strategies, especially as the country enters another very delicate stage of its history in the transition to 1992?

It is impossible to answer this question with any certainty, but it appears that several contradictory forces could emerge to strain the unity of the armed forces. The first issue concerns the demilitarization process itself. While the current military leadership is in favor of completing the job by 1992 (having already postponed this from 1990), it is rumored that some officers believe that this date is too early for any noticeable improvement in civilian behavior and political habits to have occurred. Some officers, for selfish, "nationalistic" and/or ideological (ie. socialist) reasons, favor the continuation of military influence over political life either directly or through a diarchical power-sharing relationship of civilians and military in government, an idea originally mooted in 1972 by Dr. Nnamdi Azikiwe, the country's first president. This structure was eventually rejected by Babangida for 1992, but there is to be some form of power-sharing between 1991 and 1992, when civilians will control state and local governments, while the military holds on to the federal government. Whatever takes place over the next two years, contrasting perceptions are present within the military: legitimate rulers for national redemption, or temporary caretakers in a civilian state?

A second point to note concerns the tensions being generated by the ongoing recivilianization, and by the lifting of the ban on political activities in May of 1989. Will the military be able to remain aloof, neutral and dispassionate? The answer to that question is both yes and no. It is probable that military discipline and esprit de corps will be maintained if the political process itself remains peaceful. Strong military control of, and adjudication between, civilian groups was very effective in the corresponding transitional period of 1977-79, and military cohesion and confidence appears to be holding at this time.

The real difficulty emerges when attempting to consider a possible scenario in which social cohesion breaks down under the pressures of the transitional period, and civil disorder occurs. It would still be feasible to foresee the military acting as a quasi-neutral force, but it is also possible that the military itself could become a victim of sociocultural tensions. The abortive coup in April 1990 and the subsequent execution of seventy soldiers exemplifies internal strains. Such tensions are likely to be further exacerbated by the fundamental economic problems and structural deficiencies present in Nigeria, and the extreme hardships faced by the majority of Nigerians. At such a juncture, actions by military leaders would be crucial. If the normative values of national integration instilled at staff colleges hold up to this test, then one can conclude that the military serves as a vehicle for integration -- both in terms of its own unity and as a model for society to follow.

One hopes that the military, and the country as a whole, can emerge unscathed and more united from the political transition program, and that the civilian government post-1992 will truly mark a new era in Nigerian politics. If this occurs, then the military could claim a measure of success in achieving the elusive objective of state-consolidation. Unfortunately, the omens are not good.

Notes

1. These are listed in the bibliography, but special attention should be paid to Luckham (1971), Miners (1971), Odetola (1978 and 1982), Oyediran (1979), and Panter-Brick (1970, 1978).

2. The terms West and East are used, but these regions are geographically in the south-west and the south-east of the country.

3. The country works from the 1963 census figures today. The 1973 census, undertaken by General Yakubu Gowon, was not adopted. President Babangida has committed his government to a census in 1991.

4. The author appreciates the theoretical problems which deter the use of the word "modernization".

5.   The rise of the Maitatsine fundamentalist Islamic movement in northern Nigeria after 1980 has been a serious threat to stability. It appears to be controlled for the present. Figures are rather vague, but Islam is said to account for over 75 per cent of the population in the North, and 20-30 per cent in the South.

6.   Miners (1971) suggests that Ibo officers were responsible for an attempted coup immediately after independence in October 1960 but that this was easily suppressed with the help of expatriate officers.

7.   Demobilization of three major categories took place: those retired troops who had been conscripted back into the army for the civil war, those troops over 55 years of age, and those guilty of certain offenses.

8.   For a discussion of the coup attempt and its aftermath, see New African (London), February 1986, and West Africa (London), 17 March 1986. Personal grievances appeared to be prominent.

9. The OIC issue is discussed in West Africa (London), 3 February 1986, and in New African (London), March 1986.

10.   The census exercise of 1973 had shown many signs of corruption.   Population in many areas of the North had miraculously increased by 100 per cent since 1963, while some Yoruba areas mysteriously recorded a 10 per cent decrease. The census was annulled by General Murtala Muhammed in 1975.

11.   Troops from Yakubu Gowon's ethnic group, the Anga, were also implicated in the coup, exemplifying the ethnic sentiments within the army at the time.

12.   This was made more as a gesture to the military than with any firm belief in its ability to deter military intervention. A similar clause has been inserted in the 1992 constitution.

13.   A similar, far-reaching policy was implemented for civilian students going abroad.   Various estimates of Nigeria's public debt were made in 1990, but most agreed that they were in the region of $30.5 billion.

14. For Obasanjo's perspective on this see Malcolm Fraser and Olusegun Obasanjo (1986) "What To Do About South Africa" Foreign Affairs, 65(1), Fall. Also see Commonwealth Eminent Persons Group on Southern Africa (1986) Mission to South Africa: The Commonwealth Project (Harmondsworth: Penguin).

15. Olusegun Obasanjo sent soldiers into all secondary schools between 1977-79 in an attempt to bring more discipline to school pupils.

16. For a specific discussion of the popular perception of the first period of military rule in Nigeria, see Peil and Olorunsola (1977). For a general discussion of the performance of militaries in Africa, see Ravenhill (1980).

17. As a gesture to human rights, the Buhari government decided to execute convicted women in private.

18. The 1984 decree made it an offense to publish any "false" statement about the government's policies, as well as to ridicule any public officer. Several journalists were imprisoned on the basis of this decree.

19. For a discussion of the military role of "parasitic" behavior and "structural violence" within African and Third World societies, see Wolpin (1986).

20. For an interesting discussion of the problems of demilitarization see Third World Quarterly (1985), January, 7(1), complete issue. See also Bienen (1978), pp. 252-265.

References

Adekanye, J 'Bayo. 1985. "The Politics of the Post-Military State in Africa." In Clapham and Philip, eds. 1985, pp. 64-94.

Akinwowo, Akinsola. 1977. "Military Professionalization and the Crisis of Returning Power to Civilian Regimes of West Africa." Armed Forces and Society 3 (4), pp. 643-54.

Bienen, Henry. 1975. "Transition From Military Rule. The Case of Western State Nigeria." Armed Forces and Society 1 (3), pp. 328-43.

____. 1978. Armies and Parties in Africa. New York: Africana.

Brownsberger, William N. 1983. "Development and Governmental Corruption – Materialism and Political Fragmentation in Nigeria." The Journal of Modern African Studies 21 (2), pp. 215-233.

Clapham, Christopher and George Philip. eds. 1985. The Political Dilemmas of Military Regimes. London: Croom Helm.

Decalo, Samuel. 1985. "African Personal Dictatorships." The Journal of Modern African Studies 23 (2), pp. 209-237.

Diamond, Larry. 1982. "Cleavage, Conflict and Anxiety in the Second Nigerian Republic." The Journal of Modern African Studies 20 (4), pp. 629-668.

Dudley, Billy. 1982. An Introduction to Nigerian Government and Politics. London: Macmillan.

Enloe, Cynthia. 1980. Ethnic Soldiers: State Security in Divided Societies. Athens, GA: University of Georgia Press.

Falola, Toyin and Julius Ihonvbere. 1985. The Rise and Fall of Nigeria's Second Republic 1979-84. London: Zed.

Gutteridge, William. 1965. Military Institutions and Power in the New States. New York: Praeger.

____. 1975. Military Regimes in Africa. London: Methuen.

Janowitz, Morris. 1964. The Military in the Political Development of New Nations. Chicago: University of Chicago.

____. ed. 1981. Civil-Military Relations – Regional Perspectives. Beverly Hills: Sage.

Kelleher, Catherine McArdle. 1974. Political-Military Systems: Comparative Perspectives. Beverly Hills: Sage.

Kirk-Greene, Anthony and Douglas Rimmer. 1981. Nigeria Since 1970: a Political and Economic Outline. London: Hodder and Stoughton.

Laitin, David D. 1982. "The Sharia Debate and the Origins of Nigeria's Second Republic." The Journal of Modern African Studies 20 (3), pp. 411-30.

Laitin, David D. and Drew A. Harker. 1981. "Military Rule and National Secession: Nigeria and Ethiopia." In Janowitz 1981, pp. 258-286.

Lee, J. M. 1969. African Armies and Civil Order. New York: Praeger.

Legum, Colin. ed. 1975. Africa Contemporary Record 1974-75. London: Rex Collings.

____. 1976. Africa Contemporary Record 1975-76. London: Rex Collings.

Luckham, Robin. 1971. The Nigerian Military: A Sociological Analysis of Authority and Revolt 1960-67. London. Cambridge University Press.

Miners, N. J. 1971. The Nigerian Army 1956-66. London: Methuen.

Mowoe, Isaac James, ed. 1980. The Performance of Soldiers as Governors: African Politics and the African Military. Washington: University Press of America.

Odetola, Theophilus Olatunde. 1978. Military Politics in Nigeria: Economic Development and Political Stability. New Brunswick, NJ: Transaction.

____. 1982. Military Regimes and Development: a Comparative Analysis of African States. London: George Allen and Unwin.

Ojo, Olatunde J. B. 1984. "Nigeria." In Timothy M. Shaw and Olajide Aluko. eds., The Political Economy of African Foreign Policy. Aldershot: Gower.

Olorunsola, Victor A. 1972. The Politics of Cultural Sub-Nationalism in Africa. New York: Anchor.

____. 1977. Soldiers and Power: the Development Performance of the Nigerian Military Regime. Stanford: Hoover Institution Press.

Oyediran, Oyeleye. ed. 1979. Nigerian Government and Politics Under Military Rule, 1966-79. London: Macmillan.

Panter-Brick, S. K. 1970. Nigerian Politics and Military Rule: Prelude to the Civil War. London: Athlone.

____. ed. 1978. Soldiers and Oil: the Political Transformation of Nigeria. London: Frank Cass.

Peil, Margaret. 1975. "A Civilian Appraisal of Military Rule in Nigeria." Armed Forces and Society 2 (1), pp. 34-45.

Post, Kenneth and Michael Vickers. 1973. Structure and Conflict in Nigeria: 1960-66. Madison, WI: University of Wisconsin.

Ravenhill, John. 1980. "Comparing Regime Performance in Africa: the Limitations of Cross-National Aggregate Analysis." The Journal of Modern African Studies 18 (1), pp. 99-126.

Shaw, Timothy M. 1987. "Security Redefined: Unconventional Conflict in Africa." In Stephen Wright and Janice N. Brownfoot, eds., Africa in World Politics: Changing Perspectives. London: Macmillan.

Suberu, Rotimi Timothy. 1988. "Federalism and Nigeria's Political Future: a comment." African Affairs. 87 (348), pp. 431-39.

Welch, Claude E. 1970. Soldier and State in Africa. Evanston. Northwestern University Press.

Welch, Claude E. and Arthur K. Smith. 1974. Military Role and Rule: Perspectives on Civil-Military Relations. North Scituate: Duxbury.

Wolpin, Miles D. 1986. Militarization, Internal Repression and Social Welfare in the Third World. London: Croom Helm.

9

CONCLUSIONS

Henry Dietz

This volume has examined various roles of the military as vehicles for social integration. The eight cases presented have provided fertile ground indeed for describing, analyzing, and comparing the multitude of ways in which states, military institutions, and ethnicity interrelate with one another. The cases convincingly demonstrate that the three-way overlap spelled out in the opening chapter (military-ethnic, state-military, and state-ethnic) takes a wide variety of substantive and historical forms, structures, policies, and outcomes. Given this complexity, it may well be worthwhile to distill the major findings of the individual chapters and then to offer some conclusions and speculations about further approaches to the general topic. But rather than simply summarizing each study, we shall examine what each case has to say about the three binary overlap areas, recognizing that these three are by no means water-tight compartments, but should and must be viewed as analytic conveniences. Where one's boundaries end and another's begin is unclear and ultimately not especially important.

State-Ethnic Relations

Butler's work on the United States Army portrays an institution that has made large strides over the past four decades as a vehicle for social change, both within itself and for the society at large. The military's unique structure -- its separateness from society, its strict hierarchical nature, its practice of enforced close contact between different groups, and its advancement on merit and not ascription -- allowed it to undertake rapid and successful integration, and to send its veterans back to a civilian society where they could operate at an advantage, at least when compared with their non-veteran counterparts. Overall, Butler credits an approach of "individual nationalizing" within the military as crucial in its success. The United States Army emerges as an institution that does not mirror society; rather, it is an institution that in some respects offers a model for the rest of society to emulate. Again, such a conclusion is relative. Butler does not argue that the military is perfectly integrated, but that relative to the larger society, it has made significant and basic progress. Whether one should conclude that the Army is advanced or the larger society retrograde is, of course, open to question.

Contrasts between the United States, on the one hand, and several of the other cases under consideration on the other are nothing less than stark. Roumani's study of Israel, for instance, paints a clear picture of a military and a society interacting with one another to maintain the status quo of ethnic power distribution. The data from his study of two educational facilities designed to provide compensatory education and socialization to disadvantaged recruits all point to a failure to integrate Ashkenazim and Orientals. But Roumani goes further. Content analysis of course materials reveals a consistent, stereotypical downgrading of the "backwardness" of the East and the "enlightenment" of the West. Such stereotyping is a direct carryover from the civilian society into the Army, and argues that the educational facilities examined perpetuate rather than change societal stratification norms. Roumani explains such findings on the basis of a search for security by both the military and the society: each one, dominated by Ashkenazim, ensures dominance by that group which is perceived as most trustworthy - which is, of course, the group that rules Israel and that runs the military. The state of Israel and its military therefore mirror one another, closely and purposively.

The remaining cases do not have the benefit of Roumani's detailed empirical data, and yet the question can still be addressed. In Ethiopia, a dominant minority (the Amhara) has been in control of both society and military for decades. Prior to the Mengistu revolution, Amhara Christians dominated the officer corps, and the two military training schools drew from highly distinct social classes: one recruited disadvantaged youth (largely Oromo), and the other some of the elite (largely Amhara). But the revolution, which was, after all, directed against Amharic Haile Selassie, has not meant Omoro dominance, either in society or in the military. Major revolutionary institutions such as the Derg, the COPWE, and the PMAC have all made rhetorical and policy moves toward the elimination of regional and ethnic disparities and discrimination, and yet Amharic dominance through repression, war, subordination, and cooptation persists in both the military and society at large.

Nigeria with its manifold diversities confronts many of the same general problems as Ethiopia, but clearly has its own set of difficulties. The nation is perhaps everyone's favorite example of a colony that was carved out by an imperial power with no regard at all for tribal or geographic considerations. Since independence, civilian political structures tend to mirror such cleavages; Wright points out that political parties reflect geo-ethnic differences, and that within the military, ethnic identification has always been an important factor in who rules the troops. A 1985 coup brought to power Inrahim Babangida, a native of the so-called Middle Belt. Babangida has attempted to lessen northern or Hausa-Fulani influence within the military, but such a move runs counter to national elite structures that are dominated by northerners.

An additional way in which the military and civilian sectors act in accordance with one another lies in the omnipresence of corruption within each sector and across the two. Characterized by Wright as "Nigeria's national pastime," corruption by military and civilian elites rose to great heights under the oil boom of the early 1980's, and Nigeria's populace apparently sees little difference between civilian and military regimes insofar as such practices are concerned. Indeed, Wright emphasizes that the population does not share the military's self-perception as a savior of the nation, and that while some policies aimed at integration within the military have had some partial success, ethnic identification and aptness at corruption count for far more throughout Nigerian society than does merit.

Thomas argues that overall the Indian military has been a positive force insofar as its roles as a defender of civilian rule and democracy are concerned. The military accepts civilian dominance in virtually all political matters and accepts as well its role as contributor to ethnic integration of Indian society as a whole. The military as an institution reflects the larger society in many ways, perhaps in large part because it is controlled by Hindu officers. One area in which the military has sometimes been reluctant to act has been internal security. Military officers apparently do not relish being typecast as "policemen," and when it comes to counter-insurgency operations against specific ethnic groups (e.g., the 1984 Golden Temple incident), the military has been hesitant to accept anything more than a short-term active role, primarily because they may fear repercussions within the ranks of the military itself.

The Chinese case, with its overwhelming dominance of the Han, would appear to be a straight-forward matter of society and military mirroring one another, and with such a reflection being no surprise. But while Han dominance of the military appears as expected, Bennett demonstrates that Han perceptions of minorities as inferior also carry through as well, with PLA recruitment demonstrating the disrespect of the Han majority for the nation's numerically insignificant minority groups. Thus societal cleavages and discrimination patterns hold true as well within the military.

Along another dimension, Bennett notes that both military and civilian elites at times develop regional networks of patronage and obligation, despite specific policy attempts by the PLA to undermine such activities. And regional differences have, from time to time, played major roles in the PLA. Officers from urban areas and from the more developed parts of China in general have in recent years (post-1983) shown signs of gaining salience within the military, thereby straining the decades-old image of the PLA as a primarily peasant revolutionary force. Yet if the PLA attempts to undertake what is generally labelled "professionalization" within the officer ranks, such a development will necessarily create severe tensions between the ideological need for a "people's Army" and the developmental desirability for a professional military officers corps.

In Greece and Turkey, Brown finds that the militaries of both nations serve as vehicles for social integration of the masses and for largely successful efforts at eradicating illiteracy. The militaries of both nations are also viewed (at least in part) as institutions that permit and encourage advancement based on merit and work in societies where such institutions are frequently rare. Brown provides data to argue that the officer corps for both countries is recruited from a generally lower to middle-class populace that may be either rural or urban in origin.

In the Greek case, the military and the civilian sectors resemble one another in ways that have already been discussed in some of our other cases. For instance, Brown describes a considerable level of distrust between military and civilian elites, a distrust based on dissimilar socio-economic backgrounds and on the presence of clientelistic mechanisms that the military may utilize despite resenting the need to do so. Yet one of the major factors assisting the military's image within Greek society today may be the trauma associated with the 1967 coup and the scandalous 1974 failure of the military junta to topple Makarios and to annex Cyprus. This failure, in some ways reminiscent of the failure of the Argentine military in the Malvinas/Falklands War, led to disgrace for the military and a forced return to the barracks. Civilian administrations since that time have succeeded in returning self-esteem to the Greek military while at the same time convincing it that its major role lies in military and not political affairs.

In Turkey, civilian and military sectors of society share a number of factors. In the first place, the Turkish military's involvement in the political arena has been fairly constant since World War II, and it is likely that a coup does not comprise the traumatic event that it might be in Greece or elsewhere. Coups in 1960, 1971, and most recently in 1980 all occurred because of what the military saw as the malfunctioning of democratic and parliamentary procedures; indeed, all of these coups had the support of at least substantial parts of major groups within Turkish society. Furthermore, the tenets of kemalism permeate Turkish society and are as well a major element of Turkish military training and socialization (indeed, a violation of kemalism has been used as justification for more than one coup).

At the same time, there are differences between civilian and military elites. Most cadets who enter the military academies are urban (the majority of Turks are rural) and have from lower to middle-class backgrounds. Specifically, a plurality of recruits are sons of civil servants, military, and police, all of which are elements of society dedicated to kemalism. Professional families, on the other hand, do not as a rule send their sons into military careers. Thus military men tend by and large to view politicians who deviate from kemalism and form the status quo as suspect.

Military-Ethnic Relations

The case studies in this volume convincingly demonstrate the continuing relevance of ethnic linkages vis-a-vis the social composition of military forces. Ethnicity remains a matter of salience both for designers of government policy and for minority elements; the latter's sense of separateness may well be heightened by skewed recruitment and career progression patterns favoring the dominant group and its allies.

Welch's examination of the Ethiopian military clearly illustrates the centrality of ethnic distinctions in stances adopted by dominant and subordinate groups regarding personnel matters. The Amhara community constitutes less than one-fifth of Ethiopia's population, yet traditionally has held sway over civil and military bureaucracies. However, force augmentation programs have brought about increased manpower demands which, in turn, have compelled recruitment of substantial numbers of Oromo (the Oromo community is more than twice as large as the Amhara) and other "minority" soldiers. Such individuals, typically filling company grade officer and enlisted slots, often have been assigned to combat units. In addition to human resource requirements, Amhara willingness to give Oromo troops important combat missions (including offensives against Eritrean separatists) likely rests on the perceived gratitude of Oromo peasants benefitting from land redistribution measures instituted by the Mengistu regime, and the inability of the Oromo to unite politically, thereby precluding any threat to Amhara supremacy.

While the Ethiopian military thus contains significant non-Amhara elements, Welch observes that thousands of minority youth have disregarded conscription laws, opting to join the opposition Tigray People's Liberation Front. Such individuals evidently deem the military an organization dedicated to securing Amhara socio-political dominance, rather than discharging integrative responsibilities.

Ethnic considerations have proven a core variable in decisions for structuring the Nigerian military as well. Politically dominant communities repeatedly have endeavored to translate their management of governmental affairs into a large group presence in senior military posts. These efforts began in the immediate post-independence period when the federal administration, controlled by northern (Hausa-Fulani) interests, established a quota system requiring that fifty percent of officer positions be allotted to northerners. Conversely, the Yoruba and Ibo communities (situated in western and eastern Nigeria, respectively) wished to abandon ethnic calculations in officer recruitment and instead base selection on objective measures, including academic success (Yoruba and Ibo endorsement of job-relevant selection criteria is not surprising, given the higher education levels attained by these groups; such an arrangement would lead to the disproportionate representation of Yoruba and Ibo personnel in the officer corps). More recent examples of ethnicity affecting manpower policies include President Shehu Shagari's promotion of northerners and demotion or dismissal of westerners (reflecting the preeminent status of the Hausa-Fulani in his civilian administration), and President Ibrahim Babangida's attempts to increase the influence of Middle Belt officers at the expense of northerners.

Nevertheless, the need for improved inter-ethnic relations has gained salience among many Nigerian leaders, who have come to view the military as a useful integrative tool. Thus, the early 1970's witnessed the absorption of numerous Biafran (Ibo) secessionists into the federal army. This gesture was extended solely to officers, not enlisted personnel (an unusual approach in that most state elites have tried to concentrate unreliable minorities in the enlisted ranks.)   It must be recognized, however, that Ibo officers have rarely become senior policy makers in the armed services.

India presents an extreme case of a multi-ethnic society where national recruitment and training efforts since 1948 independence have been heavily influenced by some colonial practices.   Notable among these colonial policies was the formation of "communal regiments" in the Indian Army that were composed exclusively of members of one ethnic group. Such practices continue today, and although efforts at creating integrated units have met with some success, the Indian Army shows no signs of abolishing the old practice.   In addition, pressures in favor of maintaining these communal units come from ethnic groups that have been favored in the past (especially, for example, the Sikhs), either because these communal units have been broken up and/or integrated or because other ethnic groups have been given priority.   Insofar as India's major Hindu-Muslim cleavage is concerned, it has not (perhaps surprisingly) played a major role in military affairs.   The Hindu majority apparently feels little hesitation about the fealty of Muslim soldiers and officers: during the 1965 India-Pakistan war, for example, Muslim units stationed near Pakistan were not withdrawn, and they remained markedly loyal to India.

The Ethiopian and Nigerian cases address group relations marked by reinforcing lines of social cleavage such as ethnicity, religion, and geographic region.   Roumani's study of the Israeli Defense Force (IDF) analyzes interaction between two Jewish communities: the Ashkenazim (East European and Russian in origin) and Orientals (Middle Eastern and North African in origin).   Orientals now comprise the majority of Israeli Jews; nevertheless, most political and economic power, as well as cultural influence, reside in the European community.   This imbalance is repeated in the Israeli military, where Orientals are generally confined to lower ranks (constituting more than sixty percent of conscripts and non-commissioned officers) and to less attractive career fields.

Adoption of European attitudes and lifestyles by Oriental

Jews represents a fundamental Ashkenazi objective. To reach this goal, the Ashkenazi see the IDF as an ideal value transmission channel. For example, IDF personnel frequently issue negative assessments of non-European behavior patterns to Oriental conscripts, while Ashkenazi behavior receives favorable characterizations.

The infusion of numerous Oriental Jews into the IDF has been impelled by manpower requirements as well as Ashkenazi desires to hasten assimilative processes. Indeed, such requirements have forced the IDF leadership to recruit individuals previously deemed unacceptable or inadequate. But the presence of poorly educated Orientals could diminish combat effectiveness, given increasingly sophisticated Israeli weaponry and the need for specialists to operate these systems. The institution of "affirmative action" programs, as recommended by Roumani, ultimately may ameliorate this situation and compel the IDF to live up to its self-declared role as an equalizer.

The disproportionate under-representation of Oriental Jews in the higher ranks of the IDF parallels other minorities and recent immigrants in the United States and other nations, using such indices as level of education and political and economic resource distribution. In Israel and the United States, socio-economic stratification has contributed, inter alia, to the underrepresentation of these groups in more technically demanding military occupations. In the case of the United States, the influx of substantial numbers of blacks into the armed forces, initially driven by President Truman's 1948 executive order ending the military's discriminatory policies, is noteworthy. Previously, US decision makers hesitated to admit blacks into the military. This reluctance was overcome only when wartime manpower shortages rendered the recruitment of blacks an imperative step. Among the initial measures designed to facilitate supervision of black groups were establishing enlistment quotas, assigning blacks to segregated units commanded by whites, and placing blacks in job categories posing little potential threat to state security (e.g., the quartermaster and transportation corps). Such structurally organized segregation, it should be noted, has never been a part of IDF policy or reality.

The US military today is an integrated institution, with blacks positioned throughout the rank structure. Indeed, nearly one in four US servicemen is black. Enhanced economic and social mobility prospects likely have generated enthusiasm among blacks for this career alternative.

Bennett sees the principal social cleavages in China as class, region, and "development differential." Nevertheless, the territorial concentration of minorities in strategic border areas has sensitized China's leadership to the continued significance of inter-ethnic distinctions. Military recruitment patterns reflect Beijing's interest in strengthening its control over border provinces. People's Liberation Army (PLA) recruitment drives periodically focus on minority communities, trusting that military service (among other assimilative measures) will result in a more tractable population. Yet minority recruitment programs have been suspended during periods of intense ethnic turmoil, such as the Tibetan revolt of 1959. At such times, Beijing apparently believes that the injection of refractory elements into the PLA would compromise its ability to execute both internal security and external defense missions.

There are pronounced differences in PLA treatment of Han and minority soldiers. For example, individuals from minority groups often are assigned to local militia units, where their familiarity with regional languages and cultural norms may facilitate accomplishment of unit objectives. Such an arrangement may show a subtle discrimination on the part of Han commanders, as regular PLA troops disdain militia work. In addition, minority servicemen frequently are returned to their home areas upon demobilization; this is not the case with Han personnel.

The Kurds (numbering six to eight million) are the largest minority group in Turkey. Residing in southeastern Anatolia, they differ from the majority population in several respects including communal structures and language. Brown characterizes the Kurds as the only minority able to threaten Turkish national unity. In this context, Ankara now exercises military rule over all Kurdish districts; armed clashes between Kurds and Turkish security forces are a routine occurrence. Given the nature of Kurdish interaction with governing elites, military recruitment strategies entailing underrepresentation of persons from southeastern Anatolia conform to expected patterns.

Greece as a nation-state is, as Brown notes, remarkably homogeneous, and thus severe cleavages along ethnic lines are not a problem within or for the military. The military depends upon universal conscription, but there are no efforts within the military to focus on specific groups within the country for recruiting purposes. On the contrary, class or ideological lines carry considerably more weight and importance than does ethnicity. For example, civilian politicians tend to come from elite and affluent backgrounds much more than do military officers. And up until quite recent times (i.e., the early and mid 1980's), the military quite clearly discriminated against anyone who had participated on the side of the communists in the Greek civil war, the idea being that anyone who had supported the communist cause could not be loyal to Greece. Such discrimination has changed, however, probably as much because of the passage of time as anything else.

State-Military Relations

Of all of the nations under examination here, the United States appears to have gone furthest in its integration of Blacks into the Army. Butler notes that Blacks appear on all levels of command throughout the Army and that since President Truman's Executive Order 9981 ordered equal treatment and opportunity for all personnel, the United States Army has eliminated color barriers at the formal level throughout the military establishment. While the percentages of Black officers throughout ranks does not equal the percentage of Blacks found in American society, Blacks in the Army perceive the military as more egalitarian than civilian institutions. In addition, Butler consistently argues that the Army was the first and foremost institution in the United States to take significant steps toward equal treatment, and that Blacks noticed such policies.

Any statement dealing with degrees of integration is necessarily relative since complete integration remains an ideal. Thus, comparisons across the various cases presented here are just that -- comparisons with one another rather than with some perfect ideal. Compared with the United States experience, the other nations fall somewhat short, and in Roumani's phrase, have at best attained certain levels of absorption without integration. Roumani's study of the Israel Defense Forces (IDF) paints a convincing picture of European or Ashkenazim Jews (today a minority in Israel) dominating not only the Oriental Jews in the IDF but in the state, society, and culture of Israel as well. His conclusion that "...the IDF programs, in effect, re-establish, strengthen, and perpetuate the Oriental/Ashkenazi dichotomy..." rests in turn on the explanation that the IDF reflects fairly rigid lines of stratification defined by ethnicity and all that accompanies that term.

Roumani's treatment perhaps provides the most detailed data concerning ethnic integration of any the cases in the volume. Yet all of the cases address the issue, and all find that integration within and by the military falls considerably short of promises, policies, and rhetoric. In the Ethiopian case, for example, Welsh argues that until fairly recent times (World War II) it was unimaginable that the military would undertake any policy designed to reach some rough proportional representation. The Amhara dominated Ethiopian society; they therefore dominated the military as well. Provincial autonomy prevailed over national unity. Yet since World War II and especially since the Mengistu overthrow of Haile Selassi in 1974, the Army has changed in some major respects. The Oromo now dominate the enlisted and conscript ranks, and the socialist goals of the government have made class and ethnic concerns secondary to ideological arguments and rhetoric. Yet at the same time, Amhara dominance continues among the officer corps. In like fashion, Wright's examination of Nigeria reveals that the military in large part reflects the extreme ethnic, tribal, and regional heterogeneity that characterizes Nigeria as a nation. Ibo-origin officer dominance over the largely Hausa-Fulani troops in the 1960's was disrupted during the abortive Biafra secession, during which many Ibo officers joined with their Ibo secessionists.

The case of China presents, at least in some respects, some basic differences. Despite Han dominance, Bennett points out that regional, developmental, and class distinctions do exist in China and are present in the People's Liberation Army (PLA). For instance, Bennett argues that recruitment policies reflect a lack of respect for minorities, that the PLA is seen as a largely intolerant institution by minorities, and that the PLA discourages social integration. Such findings may be somewhat surprising for an institution such as the PLA that has prided itself for years as a peasant army. Yet the post-Mao leadership has made moves toward professionalization of the Army, especially the officer corps. Whether such moves will diminish or reinforce discrimination is as yet unknown.

Insofar as Greece and Turkey are concerned, Brown's research indicates that both countries have made some considerable strides toward integration within their militaries. Such parallel advancements rest upon some national similarities: both countries are largely homogenous in terms of language, culture, and religion, facts which facilitate integration. In Greece, the army in particular today makes conscious and extensive efforts to train and to socialize its conscripts and (to a lesser extent) its officers, and all of its military units have, politically speaking, been fully integrated since the 1980's. The Turkish case presents some similarities: all levels of the military are inculcated with Kemalism (derived from the goals enunciated by Kemal Ataturk), an ideology that is omni-present not only throughout the military but Turkish society in general.

As Thomas makes clear, India is remarkable among the larger, multi-ethnic Third World nations for the degree of cooperation that exists between the state and the military and for the compliance shown by the military to civilian authority. Civilian control prevails, and the likelihood of active military intervention that might lead to a coup is remote. Part of this arrangement stems from the military's treatment by civilians. The military controls considerable resources in India, and civilian leaders have been sensitive to what the military considers its own internal corporate affairs such as promotion. The military and the state have also developed close relations due to such incidents as the 1962 India-China border war and the 1965 India-Pakistan conflict; both of these events, it should be noted, contributed to substantial feelings of national unity and purpose.

Where the military and the state have apparently sometimes clashed is over matters of internal security. The military is more than ready to provide assistance in the fact of disaster, but is less disposed to become involved in longterm developmental efforts (i.e., civic actions programs). The Indian military also has some severe doubts about the propriety of becoming involved in counter-insurgency actions, although there have clearly been times when military involvement was necessary and inevitable. The 1984 Golden Temple assault was one such, but when insurgencies have persisted over time, the military has backed political solutions.

Directions for Further Research

What conclusions can be drawn from these rich though extremely idiographic syntheses? The complex binary interaction among the three variables of the state, the military, and ethnic groups, along with their coming together in the integration nexus, all make overall patterns difficult to identify. Indeed, were such eight disparate cases to yield ready similarities, we would be left to wonder if our scheme were obfuscating reality and/or if the cases selected for analysis were (as they were intended to be) dissimilar enough to meet Prezworski and Teune's (1970) criteria to be "most different case comparisons." But assuming that the differences observed and highlighted by the cases examined here are sufficient to warrant labelling them as "most different," then what sorts of basic commonalities (if any) can in fact be discerned?

The theoretical linkages between the state, the military, and ethnicity each emphasize different elements. Existing work on ethnicity, for example, focuses for the most part on the role of the state in shaping and molding ethnic identity, while much research on the military tends to concentrate on recruitment, training, and deployment as methods of social integration. Finally, research on the state frequently examines the role of the military as an instrument of state control. The case studies in this volume provide evidence for all of the linkages that theorists have noted. For example, the chapters on Israel and Ethiopia discuss attempts by the state – through the military –

to change ethnic identity. Likewise, chapters dealing with China and the United States examine the role of recruitment, deployment, and training in the integrative policies of the military.

But when it comes to analyzing the complex interactions between and among these three variables, it is difficult to discern any single or overarching pattern. And rather than force the various cases to fit some theoretical procrustean bed, it may be more worthwhile to divide the cases into four subgroups, each of which examines a particular aspect or facet of the state-military-ethnicity -- or, in a word, the integration -- problem. The cases of the United States and the People's Republic of China provide examples of the integration of ethnic minorities into the military. The chapters on India and Israel examine the integration of ethnic majorities (Israel) or multi-ethnic groups (India) into the military. The studies of Greece and Turkey, in contrast, focus on the integration of civilian national values associated with a particular social/educational background by the military. And finally, the cases of Ethiopia and Nigeria investigate national integration efforts sponsored and controlled by military government. Each of these subgroups faces a distinct integration need or problem; each group, therefore, pursues a somewhat different strategy or emphasis.

However, excepting the first sub-group (China and the United States), the authors of the case studies in all instances identify concerns about state security and social order as the source of military integration policy. For China and the United States, Bennett and Butler suggest that values and beliefs are the core elements underscoring the rationale for integration policy. In the cases of Israel and India, ethnic divisions present a potentially severe threat to internal stability. In Israel, Roumani notes that the dominance of an ethnic minority in positions of power (and as the source of state ideology) requires the absorption of the ethnic majority into the minority's culture. In India, given the multi-ethnic make-up of the state, attention to issues of ethnicity in the military is paramount to prevent the domination of one group over all others. In both nations, furthermore, the military is a vehicle to mediate and ameliorate socio-economic ethnic divisions that could provoke internal conflict. In Greece and Turkey, Brown asserts that lack of integration of civilian and military beliefs leads to military coups as a result of military perceptions of civilian governments "deviating" from acceptable national policies. And, in Ethiopia and Nigeria the military, acting as the national government, must devise national policies of integration in the face of destabilizing ethnic cleavages that threaten the continuing survival of the state. Clearly, state fears about order and control motivate -- to a large extent -- the integration politics discussed in the case studies.

The last two groups of case studies -- Greece, Turkey, Ethiopia, and Nigeria -- provide an opportunity to examine the issue of state-military relations in face of military interventions. As noted in the introduction, if the military is merely an instrument of the state, then military coups seem improbable: military takeovers represent an attack on the state. The case studies, though, seem to suggest that survival of the state provided the rationale for military intervention. That is to say, the military -- accepting its role as guardian of state security -- acted against the government to protect the state. Obviously, in such cases, the military perceived the state to be distinct from the government. Ironically, the concepts of state and state security are given to the military, initially, through military education funded (and, at least nominally, supervised) by pre-coup civilian authorities. As Welch and Wright note, however, events such as fighting separatist movements and career interaction in the military effect military organizations and their perceptions, particularly with reference to ethnicity and state security.

Identifying certain types of integration efforts in turn suggests some ways in which further research might proceed. That is, identifying one of these specific integration needs in a society (or across a selection of societies) both allows and necessitates a clarity of vision that can highlight certain aspects of the complex and far-ranging issue of integration. While it can be argued that such narrowness might obscure or ignore variables that influence the whole integration process, social science theory may not be ready to take on the whole of integration (as that term is defined by the nexus of the state, the military, and ethnicity) just yet. Therefore, selecting specific integration needs permits identification of what are essentially middle-range theoretical or analytic focal points and concepts, which given the present state of research and the scope of this volume may be highly appropriate.

Rather than attempt here to formulate some sort of comprehensive theoretical framework (probably an unrealistic goal for such a volume), it may be prudent to suggest two courses of action insofar as future work is concerned: to identify some middle-range goals and to suggest as well some methodological approaches and issues to advance the cause of research.

For the first, one way of improving our knowledge of the role of the military might involve purposively looking at cases that focus on a specific aspect of the integration issue. For example, selecting cases in nations where an ethnically-inspired separatist movement is or has been active would allow a focused examination of this particular variant of state-society-ethnic interaction. Such a comparison might address a number of questions. What macro conditions tend to encourage such movements? What sorts of intra-military policies can be associated with separatism? What appears to be the timing of chains of events that culminate in separatist movements? What makes one especially violent while another may seek a more legalistic resolution? What sorts of society-ethnic relationships either ameliorate or exacerbate tensions that give rise to armed movements? What are the potential roles to be played by external forces? Horowitz (1985) suggests that secessionist movements vary by being either backward or advanced groups in their societies and by secessionist regions being advanced or backward as well (229-288), and generates a variety of propositions, but his empirical treatment tends to be more encyclopedic than in depth configurative.

Ethiopia is one clear case of separatism; the Eritrean separatists have been waging war for about thirty years in an effort to create an independent state for themselves. The Biafran instance in Nigeria is another, although one that failed in the attempt, and the Kurds in Turkey is (or may be) still another, as could a variety of ethnic groups in the Soviet Union. But a close comparison of these cases with one another and with some others -- for instance, Sri Lanka and

226 *Dietz*

the Tamil movement, Spain and the Basques, Chad and its Muslim population – could be most fruitful, especially when carried out through a framework designed to focus (as suggested in the introduction) on specific political groups, processes, and social categories that define and drive separatism. Such comparisons do not constitute, in one sense, the "most different systems" approach that this volume consciously set out to follow; nevertheless, and always given the understanding that separatism is the guiding commonality, cases as disparate as Spain and Eritrea must be considered as severe tests for any deductive or inductive generalizations that might be proposed.

Other similar kinds of middle-range foci could be suggested for enhancing the comprehension of military-society-ethnic relations; one of specific comparative interest would be military populist regimes and their impact on ethnic and integrative issues generally. Whatever the topic, pursuing it depends upon an ability and willingness to be sensitive to a number of methodological considerations. For example, in-depth, fresh information concerning training and socialization techniques for both officers and enlisted men in militaries in a number of countries depends upon access to what are often considered to be areas off-limits to scholarly inquiry. Often times academics are obliged to make do with secondary materials, and while these data sources can be rich indeed, they are never as complete as might be desired. Sometimes circumstances obviate first-hand investigation. Yet survey techniques, informant interviewing, and semi-structured interviews can still be used, and are especially appropriate for gathering rigorously comparative data.

In like fashion, comparative longitudinal studies that could establish baseline data on military recruits before entering the military and then follow them through their training and active service and on into civilian life and careers could be most useful, especially if the data were collected so as to be comparatively analytic. Few studies dealing with longitudinal data collection have been done outside the United States and Western Europe, and there is no gainsaying the difficulties that might be encountered in carrying them out. Yet again, innovative use of existing data sources and sensitive techniques for conducting surveys should not be discarded simply because they are untried.

Whatever the topic, the focus, the approach or the methodology, there is much to be done. The collection of essays in this volume demonstrates that a wide range of information can be gathered that is useful not only for its descriptive richness but also for its comparative utility. Given that integration (as that term is defined by the nexus of the state, the military, and ethnicity) will remain simultaneously a basic goal and stumbling block for a host of nation-states both developed and underdeveloped, both capitalist and communist, both democratic and authoritarian, the range of questions that can be asked and the sorts of frameworks that can be proposed for answering them are limited only by the individual undertaking the investigation in the first place.

Acknowledgments

I would like to acknowledge the assistance of Renee Gannon and Nancy Jeffrey, both of whom made contributions to this chapter that were of considerable analytic and substantive use.

References

Horowitz, Donald. 1985. Ethnic Groups in Conflict. Berkeley CA: University of California Press.

Prezworski, Adam, and Henry Teune. 1970. The Logic of Comparative Social Inquiry. New York: Wiley.